WILDFLOWERS
of the LAKE MEAD REGION

Common Wildflowers, Shrubs and Cacti
of Lake Mead National Recreation Area

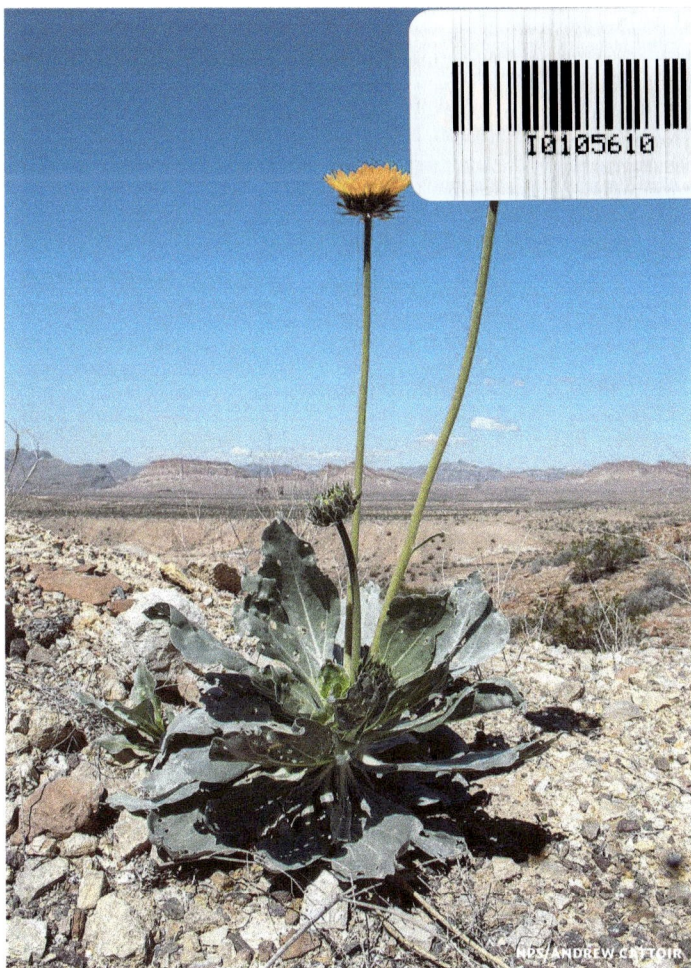

SUNRAY (*Enceliopsis argophylla*)

STEVE CHADDE

Enjoy your visit to our nation's national parks, monuments, and recreation areas, but remember, plant life is protected, and please refrain from picking any of the plants you see in these areas.

WILDFLOWERS OF THE LAKE MEAD REGION

Common Wildflowers, Shrubs and Cacti of Lake Mead National Recreation Area

STEVE CHADDE

A Pathfinder Field Guide, published by Orchard Innovations

ISBN 978-1951682675

The author can be reached at: steve@chadde.net

VER. 1 (7/25/2022)

CONTENTS

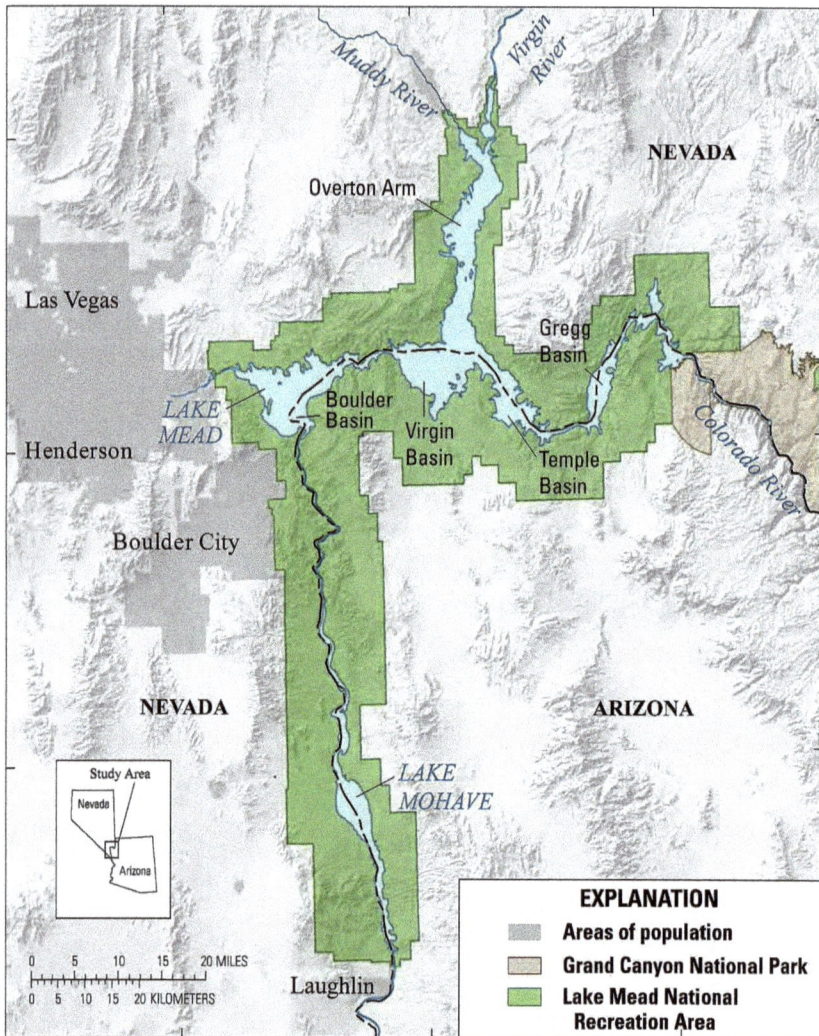

Map of Lake Mead National Recreation Area and vicinity. (USGS).

INTRODUCTION

The desert has often been described as harsh, unyielding, barren, desolate, and dangerous. At first acquaintance, the desert landscape of the Lake Mead region, located a short distance east of Las Vegas, Nevada, seems to merit any or all of these terms. It is hot during summer months, with temperatures sometimes ranging up to 120°F (49°C) or more in the low valleys.

To conceive of much life existing in the stifling heat would seem to require a stretch of the imagination. It does appear to be almost devoid of plants; certainly those reaching the size of trees.

As for that all-important life sustaining substance—water—the dry washes, slopes, and canyon bottoms always seems to have a parched look. Such an appearance is natural as only about three percent of all rain that falls on the desert penetrates the soil to any appreciable depth. This is true also of the frequent cloudbursts that strike the region during July and August. Torrents of water fall, flood the washes in a few moments, and as quickly run off, leaving little to benefit plants of the area.

It is when one gets better acquainted with the desert that its true characteristics are revealed. It is harsh, but also fragile and easily destroyed. Plants and animals living here are often literally on the thinnest edge of existence. With annual rainfall of some 4 inches (10 cm) or less, an inch of moisture can easily mean the difference between life or death for several species. The loss of a couple of inches of rainfall during the year can be a disaster if prolonged into following years.

Barren and desolate it surely appears, but this desert contains many plants and has the potential for many more under the right conditions. Animal life is widespread, involving numerous species of mammals, birds, reptiles and even some amphibians. Invertebrates, such as insects, spiders, scorpions. etc., are common. Thus one's first impression of barrenness and desolation is truly deceptive.

For given the right conditions of moisture, temperature, and soil, the desert can and does become a vast flower garden on a scale difficult to imagine. However, such conditions do not happen every year. In fact, there usually are several years after one "mega-bloom" before another comes along.

Every living thing in the desert has three basic problems to solve if it is to survive. These are: scarcity of water, high temperatures, and availability of food. In the plant world, these problems are met in a variety of ways and with amazing success.

Plant Groups

Plants of the desert region divide naturally into three well-defined groups, depending upon how they meet the problem of drought survival:

THE DROUGHT ESCAPERS

These are the annuals, the most abundant and showy of desert plants, with some of the tiniest and most exquisite flowers. They have no water problem, for if there is a lack of water, they do not germinate.

Seed germination is not a haphazard affair. Unless certain well-defined conditions are met, seeds lie dormant on the ground, usually thinly covered with the fine dust or blow sand brought in by the strong winds that are so prevalent. The seeds of most of these annuals contain a substance which acts as an "inhibitor" to germination. To dissolve away this protective material, approximately an inch of water must fall during the colder months of autumn, winter, and spring seasons. At the same time, desert winds must not be too drying and the mean temperature must remain around 65°F (18°C) or lower. The seeds will then germinate.

The reverse is true for annuals that bloom in the heat of summer. Summer rains will dissolve away the germination inhibitor if the mean temperature stays around 80°F (27°C) or more. Thus, no matter how much rain falls in winter, seeds of the summer annuals will not germinate; the reverse is true with the winter annuals when heavy rains fall in the summer. In both instances, however, the life of the desert annual is brief. With an extremely short growing period available, each plant must quickly achieve its one objective—accomplish flower pollination and produce seeds for a new generation. Thus, much of its energy is devoted to production of the all important flower.

Some flowers are designed to enhance the possibility of visits by insects and other pollinators, and this may increase the rate and quantity of seed production. Because some insects, such as the bee, can see yellow, green, blue, etc., but not red, most annuals have one or more of those colors in their petals, thus increasing the chances that they will not be overlooked. Even the red-petaled flowers, such as monkey-flower and penstemon, have yellow in their stamens. White-petaled annuals, also abundant, carry the important colors for pollination within their flower parts. Night-blooming plants depend primarily upon odor to bring pollinators to the flowers. Moths do most of this work. They do not react to color, but do have a keen sense of smell.

Once seed production is assured, the plant soon withers and dies. Seeds fall to the ground, there to be moved about by wind and rain until they finally come to rest, ready for germination at some future time. Such a time may be years in coming, but sooner or later they will have their few weeks of colorful beauty.

THE DROUGHT EVADERS

These are perennial plants living for many years, that meet the twin problems of water scarcity and high summer heat by reducing all but essential life processes. They flower in spring along with the annuals, but when summer temperatures arrive, they may shed their leaves and enter a state of dormancy until suitable growth conditions are once more present. The mariposa lily and the ocotillo are good examples.

THE DROUGHT RESISTERS

This group includes a wide range of shrubs and other woody or fibrous plants. These take the worst conditions the desert offers and manage to survive. Some, like members of the cactus family, store water in their stems or root tissues. Others, such as mesquite and catclaw, depend upon widely spreading root systems designed to utilize every available bit of moisture in the soil. Still others, like the desert senna, rely upon reduced leaf surfaces. Some plants, such as brittle-bush and desert holly have fine, gray, downy coatings on leaves and stems which reflect the sun's heat away from the plant. Leaves of the creosote-bush are coated with a waxy material that reduces moisture loss. All of these plants manage to live from one growing season to the next, adding new growth whenever opportunities are favorable.

Plant Communities

Plants, like people, have definite preferences as to where they live. Some prefer open desert. Others like the rocky slopes and cliffs, dry steambeds, or where permanent water is found. Some even live on soils containing minerals that are toxic to many other species. Plants of the Lake Mead region can be grouped into five, rather well-defined communities, illustrated below:

NPS/ANDREW CATTOIR

THE JOSHUA FOREST COMMUNITY

The Joshua-Tree is found between around 3,500-5,000 feet (1,066-1,524 m) elevation, and several other species are associated with it. At this higher elevation, rainfall is more plentiful and summer temperatures are not as high as on the desert floor. Light snows may be received in winter.

LAURA CAMP

KRISTINA HOEPPNER

THE CREOSOTE-BUSH COMMUNITY

The most conspicuous plant of the region is the creosote-bush, found in a wide elevational range of between 500-3,000 feet (152-914 m) and covering extensive areas. It is most abundant in the middle desert zone around 2,000-2,500 feet (609-762 m). Rainfall is low, only about 3-4 inches (7-10 cm) a year. Annual temperatures range from 10°-115 °F (-12° to 46°C). Other common plant species associated with this community include burro-bush, cholla, senna, and indigo-bush, to name only a few.

MARCWINGS

THE DESERT WASH COMMUNITY

This community occurs from elevations of about 500 feet (152 m) to as high as 3,000 feet (914 m). Water concentrates here during rainstorms and because more underground water is available, plants grow in greater abundance than in the surrounding area. The washes are subject to flash-flooding and are where most of the spring-flowering annuals will be found. Typical plants include cheese-weed, chuckwalla's delight, desert-mallow, and catclaw.

NPS/ANDREW CATTOIR

THE CLIFF COMMUNITY

Narrow, steep-walled canyons are often found at the upper ends of desert washes. Plants here prefer the rocky slopes and cliffs, often growing out of cracks in the rock walls. Representative species include desert-fir, rock daisy, barrel cactus, and rock-nettle. In the photo above, Brittle-bush lines the rocky ravine of Lake Mead's Majestic Canyon.

THE DESERT SPRING COMMUNITY (*not illustrated*)

Plants of this community grow around springs, along river courses and in low washes where water is found at or near the ground surface. They include cattails, rushes, arrow-weed, desert-willow, mesquite, and saltcedar (tamarisk). At Lake Mead, several springs support small populations of California Fan Palm (*Washingtonia filifera,* below).

FOREST AND KIM STARR

How to Use This Book

This book includes only plants most commonly seen in the Mohave Desert (which encompasses Lake Mead), northern reaches of the Lower Sonoran Desert (which lies south of the Mohave Desert), and, in particular, those found in the Lake Mead-Lake Mohave region. Many less common and less conspicuous plants are not included.

To make identification easier, the book is divided into sections according to primary flower color. Botanical names (scientific names) in this guide follow those of ITIS, the Integrated Taxonomic Information System (*www.itis.gov*). The botanical name of the plant, as well as that of the family to which it belongs, is given in the descriptive text.

Throughout the text, reference is made to elevations where each plant is usually found. The three elevational zones are:

» **Lower Elevations**—areas from 500-1,500 feet (157-427 m)
» **Middle Elevations**—areas from 1,500-3,000 feet (457-914 m)
» **Higher Elevations**—areas from 3,000-4,500 feet (914-1 371 m)

While there are mountains in the Lake Mead region that rise above 4,500 feet (1,371 m), no attempt is made here to include plants found at these elevations.

DICK CULBERT

KRZYSZTOF ZIARNEK

FLAT-TOPPED BUCKWHEAT

Eriogonum fasciculatum **BUCKWHEAT FAMILY**

This buckwheat, in contrast to many other members of this family, is a low evergreen shrub. The leaves are small and numerous, often with the edges rolled under. Clusters of pink to white flowers can be found atop leafless stalks during the late spring months, and may also be commonly seen in early fall if sufficient rains have fallen. These stalks will turn deep red with decreasing temperatures of autumn.

WHERE FOUND Most common in washes and canyons of the mountains throughout the region.

PATRICK ALEXANDER

PATRICK ALEXANDER

MATT LAVIN

WILD RHUBARB

Rumex hymenosepalus   **BUCKWHEAT FAMILY**

Known over much of the desert country as "dock," this is an easily recognized reddish-colored perennial. The plant has large leaves, wavy margined, and the stems are thick and possess an acidic sap. The leaf petioles are good substitutes for rhubarb in pies. Native Americans used the leaves for greens, and roasted and ate the petioles. It also furnishes food for various desert animal species.

WHERE FOUND Dry washes, and often in disturbed soil along road shoulders in the southern part of the Lake Mead region.

JIM MOREFIELD

BREWBOOKS

STAN SHEBS

SAND-VERBENA

Abronia villosa   **FOUR-O'CLOCK FAMILY**

This is one of the most spectacular and fragrant desert wildflowers. Blossoms are in clusters and produce a delicate fragrance, especially noticeable in evening and early morning. As the flowers age, they quickly lose their initial radiance and fade. Some can still be seen blooming as late as June. With sufficient rain in early fall a short flowering season may develop. The stems are often as much as 60 cm (2 feet) in length, mostly prostrate. The entire plant is covered with small sticky glands, resulting in on outer covering of sand.

WHERE FOUND It grows profusely in dune areas or where patches of blowing sand have accumulated.

STAN SHEBS

KATIA SCHULZ

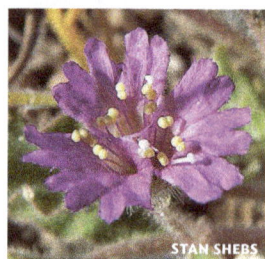
STAN SHEBS

WINDMILLS

Allionia incarnata   FOUR-O'CLOCK FAMILY

These are small, prostrate plants that spread out over the ground. Flowers are usually magenta, but may range from rose to white. There are actually three flowers in each flowerhead, which because of their arrangement, appear to be only a single blossom. Stems are sticky and hairy, often covered with find sand particles. These are perennial plants, dying back during winter months, but producing new stems and leaves each spring. They bloom from late spring to early fall.

WHERE FOUND Rolling hills and open flats throughout the region.

NPS/ANDREW CATTOIR

NPS/ROBB HANNAWACKER

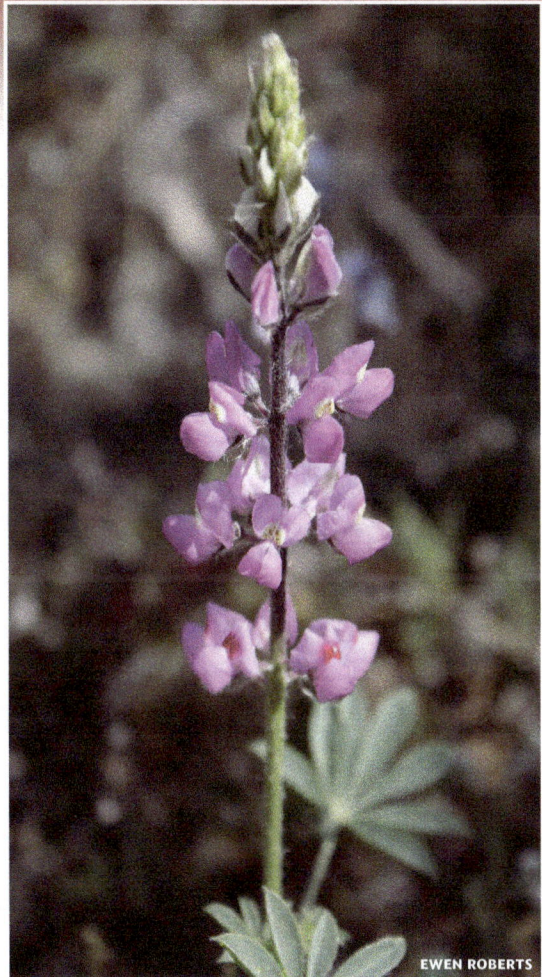
EWEN ROBERTS

ARIZONA LUPINE

Lupinus arizonicus   PEA FAMILY

This leafy-stemmed annual may grow to a height of 60 cm (2 feet), but usually much less. The leaves are compound, composed of 5 to 10 leaflets and grow along the entire stem. Their arrangement is termed palmate because the leaflets arise from a common point. Seed pods are hairy, about 2.5 cm (1 inch) in length and contain 5 to 7 seeds. The seeds are an important food source for small animals, such as mice and birds. They bloom from March to May. The Royal Desert Lupine (*Lupinus sparsiflorus*), a close relative, often grows in the same area. It differs only in having blue flowers and narrow leaflets.

WHERE FOUND Dry washes and gravel slopes, road shoulders.

ANDREAS ROCKSTEIN

ANDREAS ROCKSTEIN

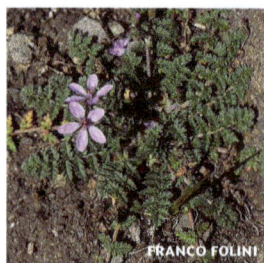

FRANCO FOLINI

FILAREE

Erodium cicutarium   **GERANIUM FAMILY**

Introduced from the Mediterranean region, filaree is a common annual throughout the western United States. The delicate and spreading stems produce numerous finely dissected leaves. Small purple flowers appear in February and persist through May. The fruit is composed of a long sterile projection with the five seeds produced at its base. Attached to each seed is a hairlike projection which is humidity sensitive. Changes in humidity cause this projection to coil and uncoil, allowing the seed to penetrate the soil surface. Populations of Gambel's quail are dependent on the seeds as a food source. **WHERE FOUND** Common along roadsides and open areas.

KRZYSZTOF ZIARNEK

SALTCEDAR

Tamarix chinensis   **TAMARIX FAMILY**

ALSO CALLED Tamarisk

SYNONYM *Tamarix parviflora*

During the late 1890s saltcedar was introduced into North America. It has spread rapidly through desert regions to become a serious problem. It thrives in a hot, humid climate and is partial to saline soils. Wherever there is water, this graceful plant forms thickets which con drain small streams or springs, thus excluding native plants and animals from the scarce water supply. In spring and early summer, pink to white blossoms form dropping colorful plumes along the shores of Black Canyon and or the mouths of washes leading into Lake Mead and Lake Mohave. The flowers attract many insects, especially honey bees and the tarantula hawk, a large, orange-winged wasp. While of little food value to other wildlife forms, it does furnish protection for birds and small animals.

EMELIE CHEN

STAN SHEBS

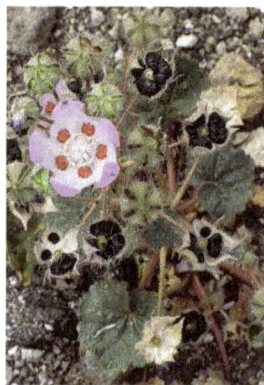

DESERT FIVE-SPOT

Eremalche rotundifolia   MALLOW FAMILY

ALSO CALLED Lantern Flower

One of the most beautiful desert annuals, the flower is round in outline and barely opens enough to show the five carmine spots on the inner bases of the petals. The plant is short, usually less than 45 cm (18 inches) high, with rounded leaves. If a hand lens is used, greenish star-shaped hairs can be seen covering the stem and leaves. It blooms from March to May.

WHERE FOUND Frequent in washes, and a sharp contrast to the somber, lava landscapes which it seems to prefer.

JAMES ST. JOHN

JOE DECRUYENAERE

OCOTILLO

Fouquieria splendens　　　　　　　　　　**OCOTILLO FAMILY**

This, one of the oddest of desert plants, is exceptionally well-adapted to withstand the harsh climate. Its long, spiny stems extend skyward as much as 6 m (20 feet) and are often completely leafless. The root system lies immediately beneath the surface, to insure maximum benefit from any moisture that falls. Small green leaves, five in a bundle, quickly cover the stems a few days after a good rain. Red flowers soon tip the branches, making a striking show. Plants lose their leaves very quickly to conserve water as soon as soil moisture becomes scarce, but will grow new leaves following more rain—this may occur more than once during the year. Ocotillo stems are often planted in rows, where they soon take root to form a spiny, living fence.

WHERE FOUND Rocky slopes; large stands may be seen in the lower Lake Mohave and Pearce Ferry areas.

CHRIS HUNKELER

STAN SHEBS

BEAVERTAIL CACTUS

Opuntia basilaris   CACTUS FAMILY

This is one of the commonest members of the Cactus family in the Lake Mead area. Its green, flat pods are covered with clumps of tiny hair-like spines. Handling this cactus can be a most discomforting experience, as the tiny spines work into the skin and are difficult to find and remove. It blooms during spring months and adds a bright spot of color to the desert landscape. Flowers soon develop into cactus 'apples'. Fruits and green pads provide food for many desert animals, especially the woodrat and jackrabbit. Native Americans also used both the fruit and the pods as food. Look carefully on the pods and you will often find green, sap-sucking stink bugs.

STRAWBERRY HEDGEHOG CACTUS

Echinocereus engelmannii **CACTUS FAMILY**

ALSO CALLED Calico Cactus, Torch Cactus

Sometimes called "calico cactus" because of the many colored spines, it grows in clumps made up of numerous stout cylindrical stems, which reach a height of 15–30 cm (6–12 inches). Flowering is from April to June. The showy blossoms close of night and reopen the following morning. Fruits are dark red, juicy, rich in sugar, and important in the diet of many birds and mammals. Native Americans of this region considered them a great delicacy. The fruit has a flavor very much like that of a ripe strawberry, hence its common name.
WHERE FOUND Rocky slopes in higher parts of the desert.

TRISTAN LOPER

MOHAVE MOUND CACTUS

Echinocereus triglochidiatus   **CACTUS FAMILY**

SYNONYM *Echinocereus mojavensis*

During May and June this clustering "hedgehog" creates a spectacular display of flowers, with as many as three flowers per stem. Each flower is 5–8 cm (2–3 inches) in length and may or may not open completely. The petals are rather thick with a waxy appearance, giving it an almost artificial look.

WHERE FOUND Rocky slopes and rock crevices in higher elevations, ranging into Pinyon-Juniper communities. In the Lake Mead area, it is found throughout the Newberry Mountains and in the Grand Wash Cliffs.

ANDREY ZHARKIKH

FISHHOOK CACTUS

Mammillaria tetrancistra   CACTUS FAMILY

ALSO CALLED Pincushion Cactus, Corkseed Cactus

Long, slender, hooked spines give this cactus the appearance of being covered with fishhooks. The plant is small, seldom growing more than 12 cm (5 inches) above the ground, so it is often overlooked. It may occur in low clumps or as a single stem. Flowers are comparatively large and are produced along the sides of the plant. Sometimes several bloom at one time, forming a crown. They mature into red, club-shaped fruits, eagerly sought by small mammals. It is sometimes called "corkseed cactus" as the seeds have a corky, brown appendage.

WHERE FOUND Look for this inconspicuous cactus among the rock-covered slopes that fan out around low-elevation mountain ranges.

JOHN GAME

STAN SHEBS

PYGMY BARREL CACTUS

Sclerocactus johnsonii   **CACTUS FAMILY**

ALSO CALLED Beehive Cactus

SYNONYMS *Echinomastus johnsonii, Neolloydia johnsonii*

Though similar in appearance to an immature solitary barrel cactus, the pygmy barrel cactus can be recognized by its round, unmarked spines. Growing to 30 cm (one foot) in height and 15 cm (6 inches) in diameter, it appears as a red-gray clump on warm, southern slopes of mountains. Wide variation in flower color makes this plant especially intriguing. In general, there seems to be a color gradient as one travels south from Las Vegas in the Muddy and Spring Mountains, the flowers take on a pinker hue; in the vicinity of Searchlight and the northern Newberry Mountains, they are a lemon yellow. Flowers remain open for a very brief time, usually for only about five days.

JOE DECRUYENAERE

STAN SHEBS

STAN SHEBS

ROCK GILIA

Gilia scopulorum   PHLOX FAMILY

This early spring annual is often found in dense stands on the gravel slopes along lower elevation washes. The flowers are small and the stems slender, so the impression one gets is that of a pink wash over a part of the landscape. The leaves are thin and rather sticky. It blooms from April to June, and in good rainfall years is especially noticeable in the washes above Willow Beach.

JOE DECRUYENAERE

DAVID D.

STAN SHEBS

FREMONT PHACELIA

Phacelia fremontii   **BORAGE FAMILY**

ALSO CALLED Yellow Throats

A common annual blooming in early spring and into May, occasionally reaching a height of 30 cm (one foot), but usually less. Leaves are basal, long and profusely lobed. The flowers are characterized by three easily recognized features: they are tube shaped, their throats are yellow, and they have a strong skunk-like odor!

WHERE FOUND Sand and gravel areas along the sides of dry watercourses, and under shrubs in the mountain valleys around 914-1066 m (3,000-5,000 feet).

NPS/ROBB HANNAWACKER

STAN SHEBS

NPS/ROBB HANNAWACKER

PURPLE MAT

Nama demissum   **BORAGE FAMILY**

The delicate stems of this beautiful "belly flower" spread closely along the ground in every direction. Near the stem tips and at the base of the plant are the flowers, usually several in number. Leaves are narrow and somewhat hairy. Several plants often grow together, forming a colorful mat of flowers, a characteristic that has given rise to the common name. It blooms from late March to May.

WHERE FOUND Sandy washes and gravels at lower elevations.

ANDREY ZHARKIKH

AREK TUSZYNSKI

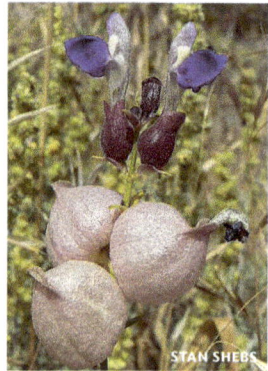

STAN SHEBS

PAPER-BAG BUSH

Salazaria mexicana   **MINT FAMILY**

ALSO CALLED Mexican Bladder-Sage, Bladder-Sage

SYNONYM *Scutellaria mexicana*

This is a low, somewhat rounded shrub with aromatic, dark green, veined leaves. The purple flowers are small and rather inconspicuous. The sepals become fused and enlarged with ripening seeds. At maturity, this fruit will break away from the plant, appearing as a small "paper bag." This large container may easily be picked up by the wind and carried a considerable distance, thus aiding in the spread of seeds across the desert. Seeds are frequently used as food by small rodents, especially the antelope ground squirrel.

WHERE FOUND Washes at middle elevations of the desert.

STAN SHEBS

STAN SHEBS

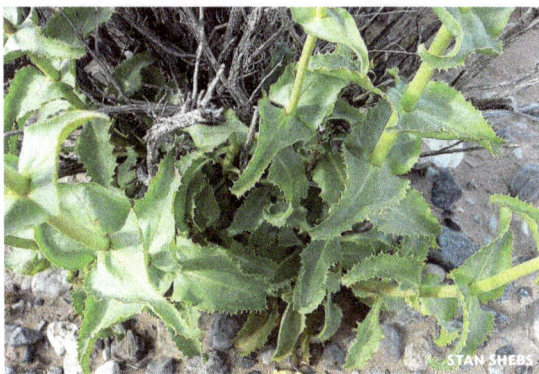

STAN SHEBS

BEARDTONGUE

Penstemon bicolor   **PLANTAIN FAMILY**

The genus *Penstemon* is large and represented by many species throughout the Southwest. This uncommon beardtongue is characterized by tubular flowers arranged along a stem 1–1.5 m (3–4 feet) high. The flowers contain five stamens, however one is usually sterile and covered with many hair-like projections—hence the common name. Leaves are opposite and fused at the base, completely enclosing the stem. A favorite of hummingbirds.

WHERE FOUND Occasional in wash gravels or Desert Wash Communities in southern Nevada and northwest Arizona.

MATT LAVIN

MATT LAVIN

JIM MOREFIELD

INDIAN PAINT-BRUSH

Castilleja chromosa   **BROOM-RAPE FAMILY**

ALSO CALLED Desert Paint-Brush

Indian paint-brush is a perennial which appears in the spring. From a woody root crown, hairy stems rise 30 cm (one foot) or more high. Leaves are narrow and about 2.5 cm (one inch) long. Flowers are barely visible, being hidden at the ends of the stem in a "brush" of scarlet-tipped, leaf-like bracts. It may grow as an isolated plant or in clumps with others of its kind. This is one plant that can be identified easily even from a moving vehicle. The bright color or attracts many hummingbirds to feed on the nectar.

WHERE FOUND Rocky crevices and dry, brushy slopes, often growing up among the shrubs themselves.

TOM HILTON

MONKEY-FLOWER

Mimulus bigelovii   **LOPSEED FAMILY**

SYNONYM *Diplacus biglovii*

This is probably the most attractive of the group sometimes humorously referred to as "belly plants." Depending on winter and spring rainfall, plants may attain a height of 20 cm (8 inches). A plant this size could easily go unnoticed, but with the relatively large flowers, this seldom happens. Flowers measure about an inch in length and are clustered near the stem tips. Stems are reddish and glandular, with thin leaves. With a short growing season before the heat of summer arrives, most of the plant's energy is directed toward producing flowers for seed production.

WHERE FOUND Gravels of desert washes at low and middle elevations.

ANDREY ZHARKIKH

SKOCH3

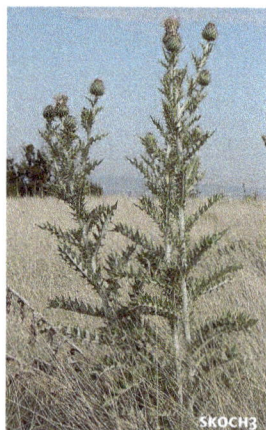

ANDREY ZHARKIKH

MOHAVE THISTLE

Cirsium neomexicanum   **SUNFLOWER FAMILY**

Thistles represent on interesting group of plants. These are biennial species. During the first year they develop a basal rosette of leaves and a root system. The following year, leafy stems and flowers are produced. After flowering, the entire plant withers and dies. The characteristic feature of thistles is the presence of spines from the elevated flowers continuing downward to the basal leaves.

WHERE FOUND Moist localities in low mountain ranges.

MATT LAVIN

STAN SHEBS

WEAKSTEM MARIPOSA

Calochortus flexuosus   **LILY FAMILY**

ALSO CALLED Straggling Mariposa

Most lilies have erect stems, but this plant has stems with a growth habit resembling that of a vine, usually growing up into low shrubs, using the woody branches for support; sometimes the stems simply wander over the ground. Flowers range in color from lavender to pink or white, and it blooms during the spring months. Like other desert lilies, the plant bulb was eaten by Native Americans. It is also a food source for small animals.

WHERE FOUND Rocky slopes and benches and, while not common, is widely distributed.

JIM MOREFIELD

BREWBOOKS

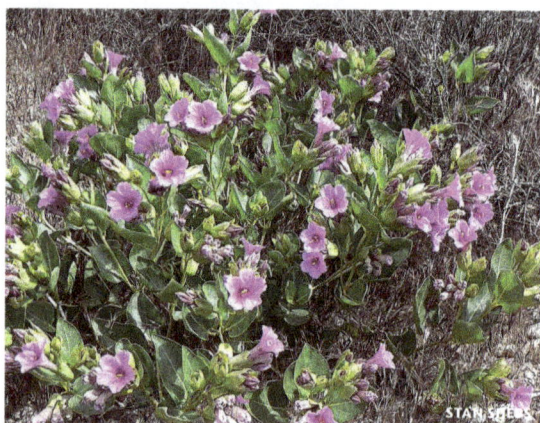

STAN SHEBS

GIANT FOUR-O'CLOCK

Mirabilis multiflora   FOUR-O'CLOCK FAMILY

SYNONYM *Mirabilis froebelii*

This desert perennial often forms large, round mats of spreading stems about a meter (several feet) across. The flowers are large, sweet-scented, and open in early evening and night. It blooms during the spring months and into early fall. The leaves are large, opposite on the stem, and unequal in size.

WHERE FOUND Open stony areas and desert washes; commonly seen along the Pearce Ferry road and also near Searchlight.

ANDREY ZHARKIKH

INDIGO-BUSH

Psorothamnus fremontii   PEA FAMILY

ALSO CALLED Mohave Dalea, Fremont Dalea

SYNONYM *Dalea fremontii*

With whitish stems, vivid flowers and gray-green leaves, this plant is one of our most attractive shrubs. It grows to a height of a meter (three or more feet) and is extensively branched. It produces many one-seeded, sharp-pointed pods, decorated with numerous small, red glands. While the normal flowering season is April and May, it may bloom again in the foil if moisture and temperature conditions are suitable.

WHERE FOUND Sandy washes and open desert.

MATT LAVIN

NPS/ANDREW CATTOIR

RANGE RATANY

Krameria erecta   PEA FAMILY

ALSO CALLED Prairie Burs, Little-Leaved Ratany

SYNONYM *Krameria parvifolia*

This low woody shrub has a dead or dying appearance much of the year. During the month of May, small green, narrow leaves appear, followed by a burst of flowers. What appear to be petals are highly colored sepals. Flower petals are minute, much shorter than the sepals. The fruit is heart-shaped, containing a single seed and covered with many delicate red spines. The roots were once dried and powdered for their medicinal
qualities in healing sores.

WHERE FOUND Common with Creosote-Bush and Bur-Sage in open desert.

JOSHUA TREE NATIONAL PARK

STAN SHEBS

STEREOGAB

TEDDYBEAR CHOLLA

Cylindropuntia bigelovii   **CACTUS FAMILY**

ALSO CALLED Bigelow Cholla, Jumping Cholla

SYNONYM *Opuntia bigelovii*

This cactus is rather tall, and its short, heavy stems are jointed and covered with a profusion of showy, silvery spines. The spines are barbed and difficult to remove, once attached to clothing or bare skin. They are deceptive and the passerby may accidentally brush against them before he or she knows it, hence the names "jumping cactus" and "jumping cholla." Flowers are either greenish or pale yellow and so inconspicuous as to be easily missed. The plant propagates itself primarily by joints on the stems, which drop to the ground and take root. The woodrat adds many of these joints to his nest, not only making his home almost immune from coyotes and other predators, but creating new plants as the joints grow. Cactus wrens build their nests amid the spiny branches.

MATT LAVIN

STAN SHEBS

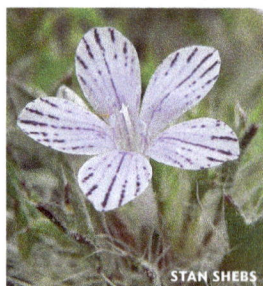

STAN SHEBS

BRISTLE GILIA

Langloisia setosissima

PHLOX FAMILY

SYNONYM *Langloisia punctata*

This annual plant is small, with stiff bristles. The flowers are light violet, with streaks of purple that are believed useful in attracting pollinating insects. Leaves are wedge-shaped, with 3 teeth or lobes at the ends. The entire plant measures a mere 7.5 cm (3 inches) across, and grows in tufts.

WHERE FOUND Found only at times of adequate rainfall; often in abundance on rolling hills and along the banks of washes at middle elevations.

MATT LAVIN

STAN SHEBS

MATT LAVIN

DEATH VALLEY PHACELIA

Phacelia vallis-mortae   **BORAGE FAMILY**

Of several kinds of phacelias in the Lake Mead region, this is the only common one that is weak stemmed and grows up among low shrubs for support. The stems are often purplish, and are covered with hair-like bristles. Leaves are long and parted into leaflets. The flowering period is March to May.

WHERE FOUND In broad, gravelly, dry washes at lower elevations.

JOE DECRUYENAERE

NOTCH-LEAFED PHACELIA

Phacelia crenulata   **BORAGE FAMILY**

ALSO CALLED Scorpion Weed, Wild Heliotrope

Two or three forms of this annual, all similar in appearance, occur in the Mohave Desert. It may attain a height of 30–60 cm (1–2 feet). Leaves are longer than wide and notched along the margins. The stems are green, glandular, and strongly scented. Sap of this plant is poisonous to many individuals and may cause severe skin rash, very similar to that produced by poison oak. It blooms profusely in the spring after good winter rains.

WHERE FOUND On gravelly, rolling hillsides and slopes.

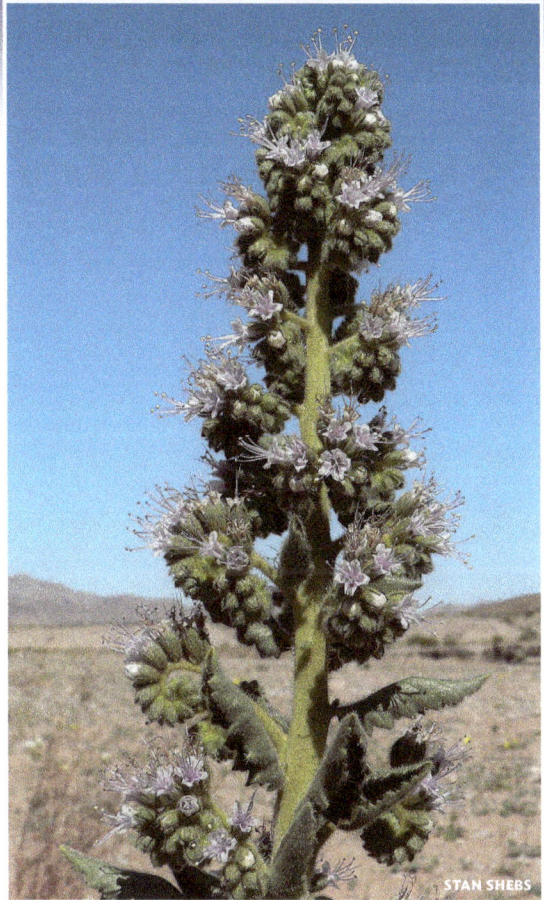

PALMER'S PHACELIA

Phacelia palmeri   **BORAGE FAMILY**

This plant may grow 60 cm (2 feet) or more in height and somewhat resembles a small Christmas tree because of the tapering growth form. It produces numerous small, rather inconspicuous blue flowers along its length. There are long, stiff hairs on the stem and shallow-toothed leaves. The entire plant is glandular and produces an extremely ill-smelling odor if handled. Although considered on annual, it requires two years to mature. It flowers in April and May.

WHERE FOUND One of the most restricted plants of the Lake Mead region, it grows only in gypsum-bearing soils, especially along the North Shore Road.

JOE DECRUYENAERE

CHIA

Salvia columbariae   **MINT FAMILY**

This annual grows to a height of 30 cm (1 foot) or more, with round clusters of flowers produced along the stem. The square stem has opposite, branching, crinkled leaves, largely confined to the base. Plants have a strong 'minty' odor. The seeds were once on important portion of the diet for many desert Native Americans. They are high in protein, and folklore has it that one can survive for an extended period on a solution of water and chia seeds. Antelope, ground squirrels, pocket mice, and small seed-eating birds also use them as food.

WHERE FOUND Chia grows on gravelly slopes and in open areas.

VAHE MARTIROSYAN

STAN SHEBS

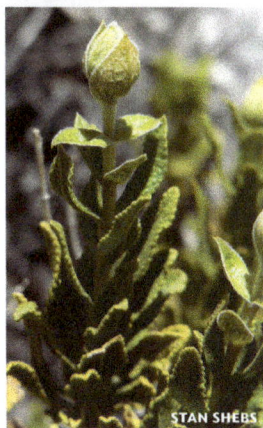

STAN SHEBS

MOHAVE SAGE

Salvia mohavensis   **MINT FAMILY**

This is a rounded, many branched, shrub growing as much as 60 cm (2 feet) in height. Leaves are dark green, deeply veined and very aromatic. The odor is quite similar to the sage used in the kitchen. It normally flowers from April to June. but when moisture conditions are right, may bloom again in early fall. **WHERE FOUND** In dry rocky washes and canyons, and is especially noticeable in the Eldorado and Newberry Mountains.

JOE DECRUYENAERE

STAN SHEBS

TOM HILTON

MOHAVE ASTER

Xylorhiza tortifolia   **SUNFLOWER FAMILY**

SYNONYM *Machaeranthera tortifolia*

This is one of the most attractive desert sunflowers. The flowers are numerous, large, and long-stemmed, so are easily observed. This perennial is somewhat woody at the base, and sends up several branches to a height of 30–60 cm (1–2 feet). The leaves are long, with spine-tipped lobes. Because of its large showy flowers, it is frequently noticed along roadsides. It blooms from March to May, and also in early fall, if sufficient rains have occurred.

WHERE FOUND In rocky areas, and is especially partial to steep-sided ravines.

ROB BERTHOLF

BLM CALIFORNIA

JOE DECRUYENAERE

AJO LILY

Hesperocallis undulata   ASPARAGUS FAMILY

ALSO CALLED Desert-Lily

This resembles a small-flowered Easter lily, and is closely related. The back of each sepal and petal has a green stripe. The flower has a pleasant fragrance. The stem is erect and 15–30 cm (6–12 inches) tall, depending upon available moisture. Leaves have a distinctive undulating margin. The bulb is deep-seated, frequently found from 30–60 cm (1–2 feet) beneath the ground surface. The name "Ajo" is Spanish for garlic. The bulb has an onion-like flavor and desert Native Americans used it as food.

WHERE FOUND A plant of blow sand and dunes. Found in the southern part of the Lake Mead region, it is most often seen along the state highway just west of Davis Dam.

MARK GUNN

JOSHUA-TREE

Yucca brevifolia   **ASPARAGUS FAMILY**

The Joshua-Tree is unique among our yuccas as it is the only one with a definite woody trunk and numerous branches. Plants are covered with short,
spine-tipped, narrow leaves. Flowers appear in dense clusters at the ends of
branches. The plant does not bloom every year, but is dependent upon availability of sufficient moisture and suitable temperatures. Average height of the
tree is 6 m (20 feet). These trees furnish important nesting sites for resident
birds, and least one species of lizard lives most of its life under the fallen
trunks and branches.

WHERE FOUND The Joshuas often form large forests in the Mohave Desert,
with an outstanding growth on the road to Pearce Ferry in the recreation
area.

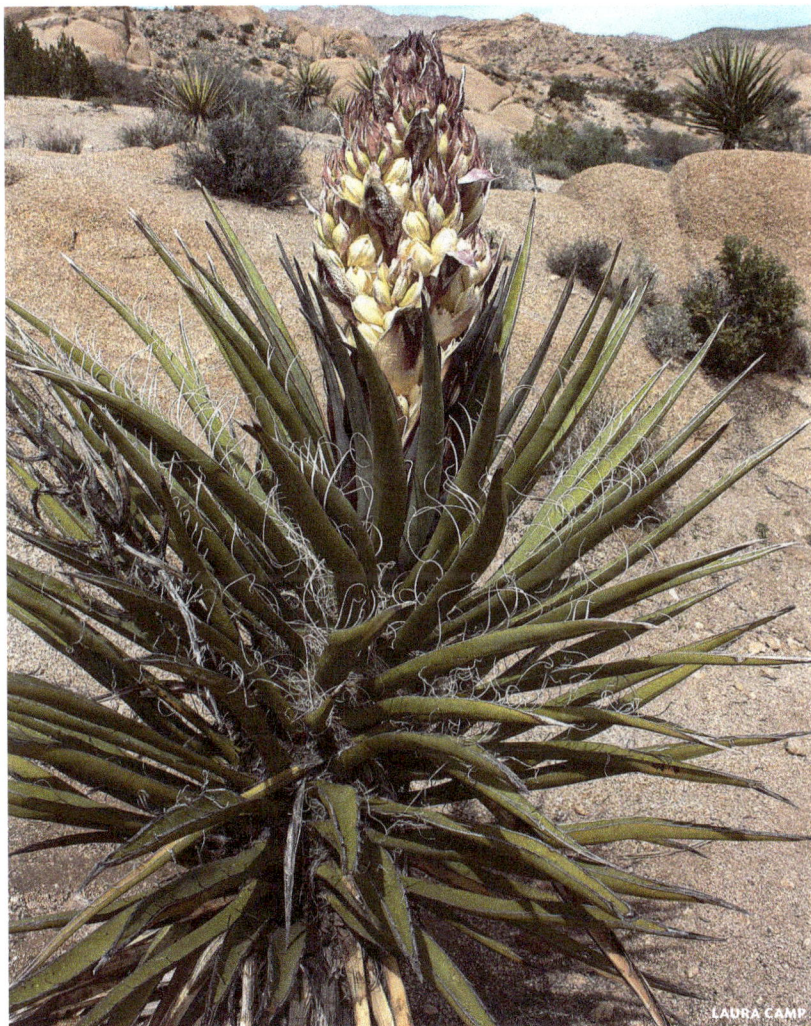

LAURA CAMP

MOHAVE YUCCA

Yucca schidigera   **ASPARAGUS FAMILY**

This is our most common yucca. It, and the Joshua-Tree, have much of their
distribution within the Mohave Desert, and are therefore considered indicator plants for this geographical region. Plants have a clumped appearance and
may reach a height of 2.5 m (8 feet). Leaves are numerous and as much as
1.25 m (4 feet) in length. They are fibrous and spine-tipped. Native Americans used the flower petals as food and ground the seeds into a fine meal.
They peeled the long, white fibers from leaf margins for use in weaving and
making rope. Leaves were also made into sandals. The seeds comprise an important source of food for several species of small mammals and birds.

STAN SHEBS

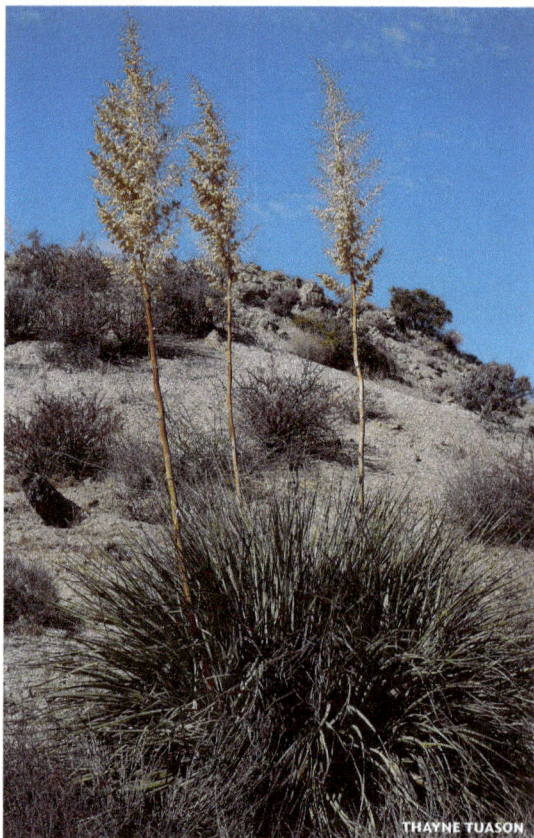

THAYNE TUASON

NOLINA

Nolina bigelovii   **ASPARAGUS FAMILY**

ALSO CALLED Bear-Grass

The Nolina is often confused with the yuccas, but its leaves are finer, grass-like in appearance and have minutely saw-toothed edges. The plume-like flowers are much smaller and less showy than those of yuccas, and appear in May. The flowering stalk, usually 1 m (3 feet) in height, is persistent and often remains intact until the following spring. Seeds are small and hard, and used as food by small rodents. Its large clump of leaves affords protection for insects, small mammals and lizards.

WHERE FOUND The plant does not grow in open areas, but prefers sheltered, rocky locations. Nolina may be found on protected slopes in the vicinity of Christmas Tree Pass in the Newberry Mountains.

PATRICK ALEXANDER

STAN SHEBS

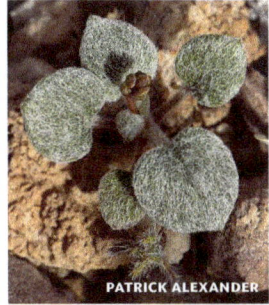

PATRICK ALEXANDER

SKELETON WEED

Eriogonum deflexum   **BUCKWHEAT FAMILY**

Over 100 species of *Eriogonum* are found throughout the Southwest, and identification of this group of buckwheats can be frustrating. TSkeleton weed is a spreading plant, forming an umbrella about 25 cm (10 inches) above the gravels in which they are found. Leaves occur only at the base of the single stem. Scattered along the horizontal branches, small pendulous white or pink flowers about the size of rice grains can be found through spring and summer. In late summer and early fall these plants take on a reddish tinge.

WHERE FOUND Fields of Skeleton Weed occur in low-lying, rocky areas at all elevations in the Lake Mead area, and are quite common along roadsides.

JOHN RUSK

STAN SHEBS

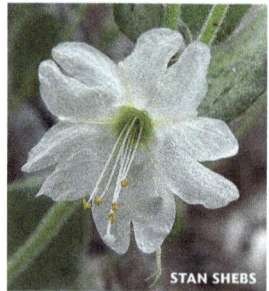

STAN SHEBS

WISHBONE BUSH

Mirabilis laevis   FOUR-O'CLOCK FAMILY

SYNONYM *Mirabilis bigelovii*

The opposite branching of this *Mirabilis* resembles the pattern of a wishbone, hence the common name. Flowers of this plant are somewhat confusing. There are no petals—only sepals—placed inside a leafy involucre. The sepals are white in color and there is a single flower in the cup-shaped involucre. To those unfamiliar with this family, there will appear to be five fused sepals and five fused petals. Watch for leaves which are opposite and usually unequal in size. The leaf character is important because few desert herbs have opposite leaves.

WHERE FOUND Look for Wishbone Bush at all elevations within the recreation area, normally on rocky hillsides.

PRICKLY POPPY

Argemone munita   **POPPY FAMILY**

ALSO CALLED Cowboy's Fried Egg

This is one of the large, showy poppies of the region. It grows to a height of 60–90 cm (2–3 feet). The leaves are lobed, and both leaves and stems are covered with needle-like, straw-colored spines which tend to prevent desert animals from eating the fleshy tissue. Stems contain a yellowish-orange sap. Large petals surround numerous yellow stamens, making the appearance of the flower similar to that of a fried egg—thus giving rise to one of the plant's common names. The seed capsule is about 2.5 cm (one inch) long and quite spiny.

WHERE FOUND Plants are commonly found from the lower elevations into the Joshua-Tree Community around 1,060 m (3,500 feet), often growing on roadside gravels and in sandy washes.

ANDREY ZHARKIKH

STAN SHEBS

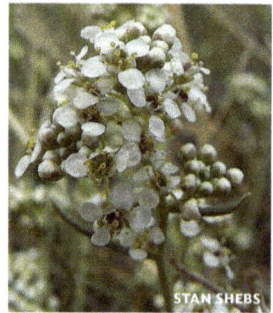
STAN SHEBS

DESERT ALYSSUM

Lepidium fremontii   **MUSTARD FAMILY**

The stems of this much-branched perennial may extend to a height of 60 cm (2 feet) and are spreading, giving plants a rounded appearance. Leaves are divided and thread-like, growing throughout the length of the stems. The foliage has a sharp and peppery taste. Flowers have tiny petals measuring less than 3 mm (1/8 inch) in length and are fragrant. The fruiting pods are notched at the tip, and each comportment contains a single seed. It was once used as a treatment for skin disorders.

WHERE FOUND Desert Alyssum prefers gypsum soils and is widespread in the region, being especially common on roadside gravels.

STAN SHEBS

NPS/BRAD SUTTON

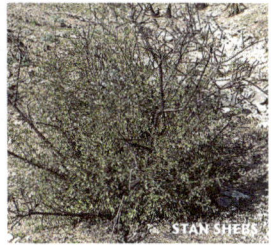

STAN SHEBS

DESERT-ALMOND

Prunus fasciculata   ROSE FAMILY

The desert-almond is a common shrub with leaves less than 2.5 cm (1 inch) in length, clustered at sites along its branches. The small, rather inconspicuous flowers appear in April. They rapidly dry and fall along with the leaves at the onset of summer. At times, large web masses of the Great Basin tent caterpillar can be observed tangled in the gray branches.

WHERE FOUND Desert-almond requires large quantities of water and is, therefore, limited to the Desert Wash Community at middle elevations.

WHITE-MARGINED SPURGE

Euphorbia albomarginata   **SPURGE FAMILY**

ALSO CALLED Rattlesnake Weed

Lying flat against the ground, white margined spurge forms green mats that may measure 30 cm (1 foot) or more in diameter. The mat is composed of many tangled branches with numerous small leaves. Stems are filled with a white milky sap. What seem to be white flowers nestled among the leaves are actually only petal-like appendages, giving on excellent imitation of a real flower. The male and female flowers are set inside and are so small a hand lens is required to observe them.

WHERE FOUND along roadsides and in gravel washes at all elevations in the desert.

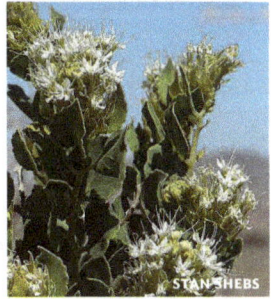

PARRY SANDPAPER-PLANT

Petalonyx parryi   **BLAZINGSTAR FAMILY**

This small woody shrub is well named, as its green stems and leaves are covered with short, barbed hairs having the rough texture of sandpaper. The plant is very stiff and the stems break easily. Flowers are fragrant and appear in profusion in the summer.

WHERE FOUND Limited to very alkaline soils, but may be seen along the North Shore Road from Las Vegas Wash to Overton.

NPS/ANDREW CATTOIR

STAN SHEBS

DUNE PRIMROSE

Oenothera deltoides   **EVENING-PRIMROSE FAMILY**

This is one of the most fragrant of desert wildflowers. The showy white blossoms are a conspicuous part of the spring flower display, but often turn pink before wilting. The flowers bloom in April and May, and open at night. Stems may extend along the ground as much as 50 cm (20 inches). Upon drying the stems curl upward and inward, forming what is often termed "baskets" or "bird cages." Other white-flowered primroses occur in the region, very similar in appearance to this species, but Dune-Primrose is the most common. Its leaves are a primary food for larvae of the two-lined sphinx moth.

WHERE FOUND Mainly in sandy soils, the plant may completely cover sheltered areas where blown sand finally settles.

STAN SHEBS

LAURA CAMP

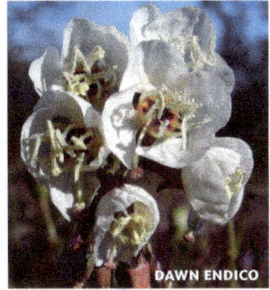

DAWN ENDICO

BROWN-EYED PRIMROSE

Chylismia claviformis   EVENING-PRIMROSE FAMILY

SYNONYM *Camissonia claviformis*

This annual is a simple, single, reddish-stemmed primrose. Leaves are green, mostly basal and toothed. The flowers may show a variety of colors, ranging from white to cream or buff. Normally there is a dark spot at the base of each petal, although sometimes this is absent. The seed capsules are on stalks, about 2.5 cm (1 inch) long and club-shaped. Plants bloom from March to May.

WHERE FOUND Common on open gravelly soils at low and middle elevations.

STAN SHEBS

PATRICK ALEXANDER

SMALL-LEAVED AMSONIA

Amsonia tomentosa   **DOGBANE FAMILY**

SYNONYM *Amsonia brevifolia*

This many-stemmed perennial grows only to about 30 cm (1 foot) or slightly more in height. The flowers, bluish when first open, soon fade and become whitish. They are borne as clusters at the tips of leafy stems. Seed pods are 5–7.5 cm (2–3 inches) in length. Stems are filled with milky sap believed to be poisonous. First flowering in the spring, plants may bloom again in early autumn if sufficient rain has fallen.

WHERE FOUND On gravelly soils above the bottoms of dry washes, usually above 900 m (3,000 feet) in elevation.

KRZYSZTOF ZIARNEK

RUSH MILKWEED

Asclepias subulata   DOGBANE FAMILY

ALSO CALLED Ajamete

Rising as much as 1.2 m (4 feet) or more above the gravel of a desert wash, stems of this plant appear as lifeless brown stalks during most of the year. With the coming of spring, the stems turn green, but do not produce leaves. Flowers soon appear on the extreme upper portions of the stems. Following an extended flowering season, long seed pods appear. These measure up to 10 cm (4 inches) in length, and are filled with flattened seeds with dense patches of silky-white hair attached to their apex.

WHERE FOUND Usually below 600 m (2,000 feet) elevation.

STAN STEBS

STAN STEBS

HUMBLE GILIA

Linanthus demissus   **PHLOX FAMILY**

This small annual is well-named as it is very low growing and easy to over-look; the entire plant is not over 5 cm (2 inches) tall. It is leafy and covered with small hairs. The petals of the flower are twisted, and each has on its in-ner surface two conspicuous purple parallel lines at its base. Flowers are fra-grant, with usually several on a plant.

WHERE FOUND Prefers gravel washes and desert slopes at lower elevations, and while often found in localized clumps, it is not common.

PATRICK ALEXANDER

JIM MOREFIELD

CHRIS ENGLISH

FORGET-ME-NOT

Cryptantha angustifolia   **BORAGE FAMILY**

SYNONYMS *Eremocarya angustifolia, Johnstonella angustifolia*
Several species of *Cryptantha* in this region have small white flowers and four-parted nut-like fruit. All have a close resemblance to one another, and a thorough examination of the fruits is required to separate these plants at the species level. All have flowers with petals fused at the base which barely exceed the sepals, and bloom in early spring. These plants are mostly bristly annuals with narrow leaves. The various kinds range in size from 8 cm (3 inches) to about 30 cm (1 foot) tall. Some are so tiny they easily merit the name of "belly flowers."

WHERE FOUND In the Creosote-Bush Community throughout the region.

STAN SHEBS

CLINTON AND CHARLES ROBERTSON

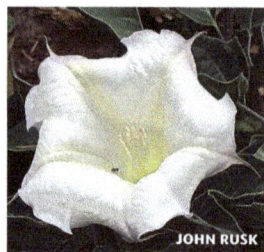
JOHN RUSK

SACRED DATURA

Datura wrightii   NIGHTSHADE FAMILY

ALSO CALLED Western Jimson-Weed, Thorn-Apple

SYNONYM *Datura meteloides*

This plant grows up to 1 m (3 feet) in height and may cover over 15 sq. m (50 square feet) of ground. The large, gray-green leaves, and trumpet-shaped flowers are a common sight along some roads and in washes. The flower blooms mostly at night, usually closing up in daytime, but may sometimes remain open if not in intense sunlight. Just before sunset it gives off a strong, sweet fragrance that serves to attract sphinx moths and thus insure pollination. Seed pods are round and thorny. All parts of the plant contain atropine and related alkaloids, poisonous compounds that have been known to cause death when eaten. Native Americans once used the plant to induce visions. It is often called "jimson-weed," the common name given its close relative by early settlers at Jamestown, Virginia.

DESERT TOBACCO

Nicotiana obtusifolia   **NIGHTSHADE FAMILY**

SYNONYM *Nicotiana trigonophylla*

This plant is usually under 60 cm (2 feet) tall, with tubular flowers about 2.5 cm (1 inch) in length. The leaves are dark green, clasping at their base. If crushed, the foliage gives off a strong, somewhat ill-smelling odor. It blooms from early spring to June, but may produce flowers into fall. It is known to contain nicotine and was smoked by Native Americans of this region in ceremonial celebrations. Another member of this family—**Tree Tobacco** (*Nicotiana glauca*)—is a small tree and grows along the shores of Lake Mead. It has tubular yellow flowers.

WHERE FOUND Prefers washes and rocky areas, especially around a cliff base.

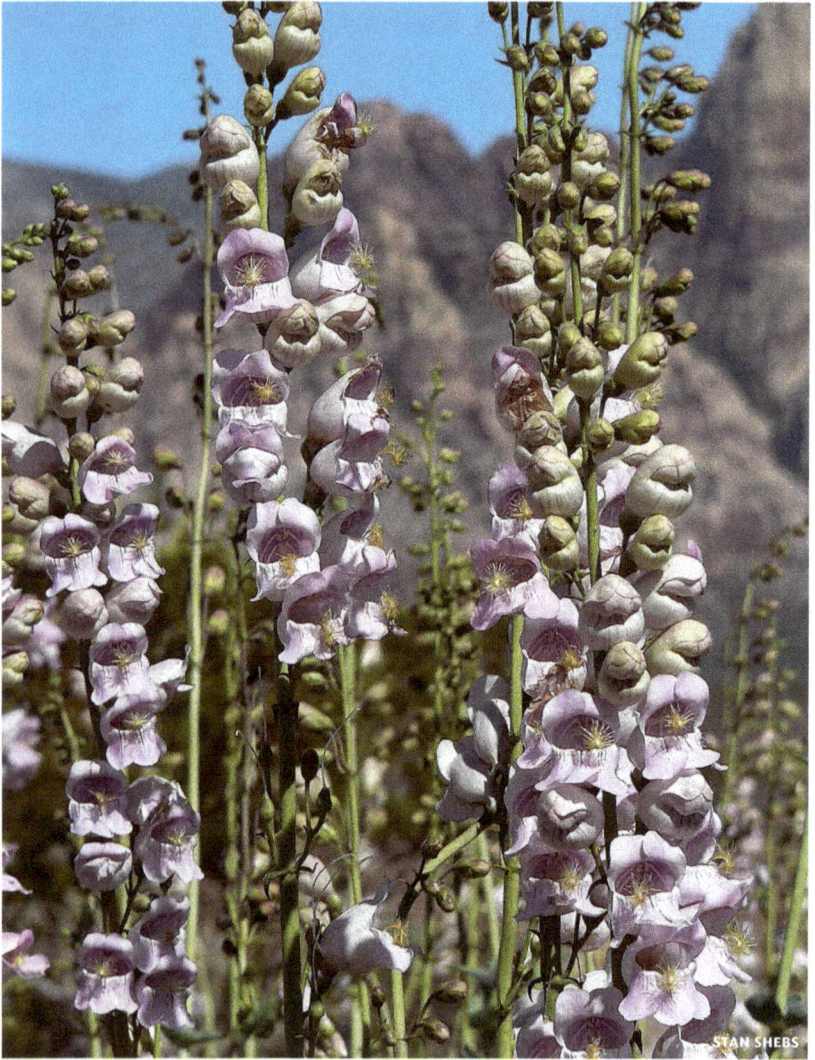

STAN SHEBS

PALMER PENSTEMON

Penstemon palmeri   **PLANTAIN FAMILY**

This is a tall plant, ranging from 0.6–1.8 m (2–6 feet) high. Leaves are sharply toothed, but do not extend far up the stem. The upper 20-50 cm (8-20 inches) of the stem bears numerous tubular white to pinkish flowers, characterized by being strongly two-lipped. Prominent purplish lines extend from the lower lips into the flower throat. The flowers produce a sweet scented fragrance (one of the few fragrant penstemons), and their nectar is much sought by bees. Leaves are used by wildlife for food, especially by deer.

WHERE FOUND In broad, gravelly washes and canyons, especially on limestone soils.

NPS/ROBB HANNAWACKER

STAN SHEBS

DESERT-WILLOW

Chilopsis linearis   TRUMPET-CREEPER FAMILY

ALSO CALLED Desert-Catalpa

This is not a willow of all, but a member of the catalpa family; its willow-like foliage has given this small tree or large shrub its name. Its trunk is crooked and has black bark. It flowers from April to July, but most of the year is inconspicuous among the heavier growth of trees and shrubs which often form thickets in desert washes. Long, slender seed pods identify the tree long after flowers are gone. The white to pink, orchid-like flowers are of such beauty that the plant is sometimes cultivated as on ornamental. The durable wood was prized for fence posts by ranchers and settlers. Leaves are used as food by larvae of the white-winged moth, and pupa cases, about 2.5 cm (1 inch) in length, may frequently be seen on the branches in late spring or early summer.

DESERT STAR

Monoptilon bellioides   **SUNFLOWER FAMILY**

This little annual easily qualifies as a "belly flower," as its stems cling close to the ground, and its dull green hairy leaves form a circle of some 10–15 cm (4 –6 inches) across. The flowering heads are found at the tips of the stems. It has the typical 'daisy' appearance. They bloom from late March to May.

WHERE FOUND Desert Star grows in rocky or sandy soils, and plants are often so dense that it is difficult to avoid stepping on them.

JIM MOREFIELD

STAN SHEBS

DCRJSR

PEBBLE PINCUSHION

Chaenactis carphoclinia   SUNFLOWER FAMILY

This annual has a rather brief life spun, as its short roots are not long enough to reach much moisture in the soil. If the spring is sufficiently wet, however, the plant grows to a foot or more in height, and produces a profusion of flowers. If moisture is limited, it develops short stems and few flower heads. The heads are rounded, and have narrow florets which appear to be "pinned" to the receptacle. Another closely related species, the
Fremont Pincushion (*Chaenactis fremontii*), is often found growing in the same general area.

WHERE FOUND Dry flats and rocky slopes at lower elevations, often growing in areas almost devoid of other plant life.

JOE DECRUYENAERE

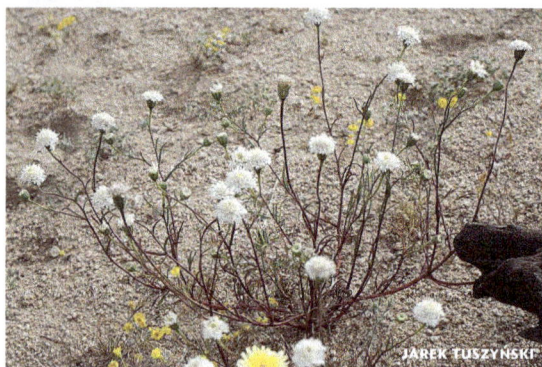

JAREK TUSZYNSKI

FREMONT PINCUSHION

Chaenactis fremontii   SUNFLOWER FAMILY

This little annual is one of the most abundant flowers to be found in a good flower year. It may have a simple stem or branch, and grows to a height of as much as 40 cm (15 inches). The stem is green, with leaves somewhat fleshy and rounded. The flowers in the center of the head are minute and five-lobed, with the stigma lobes appearing as two white "horns." Flowers around the outer margin of the flower head are larger and irregular in shape. (See also the Pebble Pincushion).

WHERE FOUND Prefers sandy soils in gravelly washes and open slopes at middle and low elevations. It often grows around the bases of Creosote-Bushes.

STAN SHEBS

STAN SHEBS

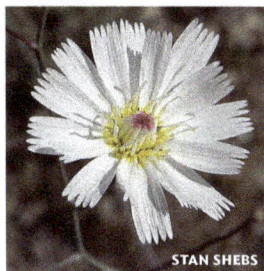

STAN SHEBS

GRAVEL GHOST

Atrichoseris platyphylla   SUNFLOWER FAMILY

ALSO CALLED Tobacco-Weed, Parachute Plant

The large flowers, borne on a tall, slender, inconspicuous, leafless stem seem to be floating in mid-air, giving rise to its common name, and also to that of "parachute plant," by which it is sometimes known. Its large, spotted, leathery leaves lie flat on the ground, and bear some resemblance to tobacco leaves. The ray flowers are sometimes purplish tipped.

WHERE FOUND Gravel Ghost prefers rocky slopes and desert washes, from low to middle elevations. It blooms from April to June, and in a good flower season can be seen in the washes above Willow Beach.

PATRICK ALEXANDER

PATRICK ALEXANDER

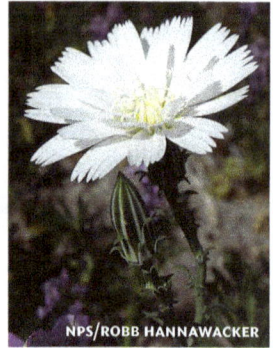

NPS/ROBB HANNAWACKER

DESERT-CHICORY

Rafinesquia neomexicana   **SUNFLOWER FAMILY**

The chicory is a weak, fragile-stemmed annual that is common when a good spring flower season occurs. The floral arrangement is very similar to that of the common dandelion. The ray flowers are veined with rose-purple on the underside, and the flower has a pleasant fragrance. It is often found growing under protective cover of shrubs, where shading tends to reduce moisture loss. It blooms from April to June. The flower closely resembles the White Tackstem (*Calycoseris wrightii*).

WHERE FOUND In washes and on gravelly slopes at middle to low elevations.

STAN SHEBS

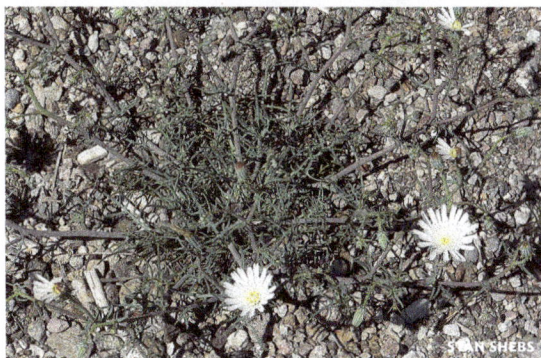

STAN SHEBS

WHITE TACKSTEM

Calycoseris wrightii   **SUNFLOWER FAMILY**

Closely resembling Desert-Chicory, White Tackstem is similarly veined,
streaked or dotted with rose or purple on the backs of the ray flowers. However, there are significant differences in stems and leaves. This plant has slender leaves, much divided; the Desert-Chicory has broad leaves, mostly at the
base of the plant, with small leaf-like bracts along stems. The stem below the
flowering head of White Tackstem has dozens of minute tack-shaped glands.
Another species, **Perry Tackstem** (*Calycoseris parryi*), has yellow flowers.

JIM MOREFIELD

JIM MOREFIELD

DESERT MARIPOSA

Calochortus kennedyi   LILY FAMILY

Under favorable climatic conditions, this short-stemmed, almost leafless lily produces a variety of flower colors; in our region the color varies between yellow and orange. At the base of each petal and sepal is a black or purple gland which helps attract pollinating insects. The plant survives the harsh desert climate by forming its bulb as deep as 60 cm (2 feet) below the surface. The bulb was used as a source of food by Native Americans and early settlers. It is sometimes dug up and eaten by small rodents. A red-flowered variety is found in Death Valley National Park. "Mariposa" is Spanish for butterfly.

STAN SHEBS

JOE DECRUYENAERE

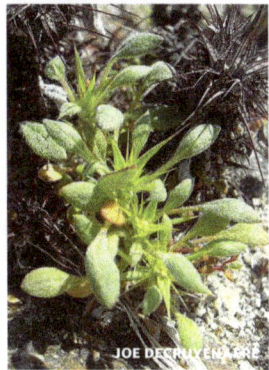
JOE DECRUYENAERE

SPINY CHORIZANTHE

Chorizanthe rigida   BUCKWHEAT FAMILY

ALSO CALLED Rigid Spiny-Herb

This is one plant which lends emphasis to the admonition "watch where you walk in the desert!" It is small, growing only 7 or so cm (a few inches) high. The leaves are woolly, and grow around the base of a cluster of green spiny bracts. Flowers are tiny and almost concealed by the spines. It blooms during spring and early summer months. The stem and spines soon become woody after the plant dies. During this time, its sharp spines pose a threat to anything walking across the desert.

WHERE FOUND Well-equipped to withstand extreme heat and commonly found in coarse gravels and areas of desert pavement at open, low elevations.

KENRAIZ

AMY WASHUTA

ANDREY ZHARKIKH

ANDREY ZHARKIKH

DESERT-TRUMPET

Eriogonum inflatum   **BUCKWHEAT FAMILY**

This remarkable plant was discovered by John C. Fremont on his journey across the Mohave Desert in 1844. The leaves form a flat mat against the ground. The leafless, branching, flower stem rises 30 cm (1 foot) or more above the mat of leaves. The plant has many uses. Stem tips taste a bit sour, but can be used in salads. Desert animals, such as the bighorn, find it an excellent food. The hollow, inflated stem is sometimes used by a tiny wasp as a rearing chamber. Drilling a small hole near the top of the inflation, she inserts several insect larvae into the cavity and lays her eggs on them. When the eggs hatch, her offspring have a ready source of food. Look for a small hole near the top of the swollen stems; if you find one, it was probably made by the wasp.

WHERE FOUND Washes and areas of disturbed soil, especially along road shoulders.

STAN SHEBS

BEAR-PAW POPPY

Arctomecon californica   **POPPY FAMILY**

ALSO CALLED Bear-Poppy, Desert-Poppy, Bear-Claw Poppy

The common name comes from the densely hairy leaves, shaped somewhat like a three-toed "foot." Several flowers form on the tip of a leafless, but hairy stem. Because it grows only on harsh alkaline soils it is restricted in numbers, this plant is being considered for placement on the federal threatened and endangered species list. Unfortunately, many flowers are picked by persons unaware that the flowers wilt very quickly, and that picking them destroys the plant's ability to produce seed. It blooms in late March and April, and is best seen near Las Vegas Wash and Rogers Spring. A white-blossomed variety of this plant occurs in Death Valley National Park.

NPS/ANDREW CATTOIR

DESERT GOLD-POPPY

Eschscholzia glyptosperma   POPPY FAMILY

This, a smaller duplicate of the California State flower, has bright yellow flowers compared to the orange-yellow of its California relative. It stands approximately 30 cm (1 foot) tall, and produces numerous blooms, each one single on an erect, naked stem. Occasionally flowers are found with pronounced orange spots at the bases of the petals. Leaves are located near the base of the plant. It produces great numbers of seeds which are used as food by small wild creatures.

WHERE FOUND Often quite abundant at lower elevations when there has been sufficient winter moisture, preferring the sands and gravels of washes and rolling hills.

CURTIS CLARK

LITTLE GOLD-POPPY

Eschscholzia minutiflora   **POPPY FAMILY**

Resembling the Desert Gold-Poppy, this little plant differs mainly in flower size. The entire plant may be only 5–7 cm (2–3 inches) across, while the flower stems bear one or more 1 cm (1/2 inch) wide blossoms.

WHERE FOUND Little Gold-Poppy grows best in broad, gravelly washes and on open slopes. In good flower years, it is common from the Boulder Beach area southward, blooming in March and April.

MATT LAVIN

PRINCE'S PLUME

Stanleya pinnata   MUSTARD FAMILY

ALSO CALLED Desert-Plume

One of the more noticeable plants of the Mohave Desert, Prince's Plume stands over 1 m (3 feet) high. The silvery-green leaves are produced mostly on lower portions of stems and are deeply cleft. Flowers are large and showy, a situation uncommon in plants of the Mustard family. Seed pods are narrow, measuring up to 7.5 cm (3 inches) in length. These plants are considered poisonous because of their ability to fix selenium from alkaline soils into the stem and leaves.

WHERE FOUND Desert Wash Communities.

STAN SHEBS

ANNE REEVES

BEAD-POD

Physaria tenella   **MUSTARD FAMILY**

SYNONYM *Lesquerella tenella*

Covering the desert with large areas of color, this annual is commonly found during spring months at low to middle elevations. It grows to a height of about 30 cm (1 foot) or more, and has several slender stems on each plant, often forming a large clump. The petals are short, only 6 mm (1/4 inch) in length. Fruits resemble small rounded beads and are on slender stalks branching from the stem. Like other mustards, it has a sharp taste.

WHERE FOUND It grows best in sandy areas.

JIM MOREFIELD

PATRICK ALEXANDER

NPS/ROBB HANNAWACKER

BLACKBUSH

Coleogyne ramosissima   **ROSE FAMILY**

The Blackbush is on important plant community indicator, growing just above the upper limit of Creosote-Bush. It often grows in such pure stands as to give a blue-gray or pale purple appearance to wide areas on the benches and slopes of desert ranges. It ordinarily grows to a height of about 60 cm (2 feet), and its branches are tough and rigid. The flowers do not have petals— sepals furnish the color. This plant, as well as Creosote-Bush, emits chemicals which may inhibit growth of competing plants. An important ground cover plant, it provides both food and protection for small forms of animal life.

KRZYSZTOF ZIARNEK

STAN SHEBS

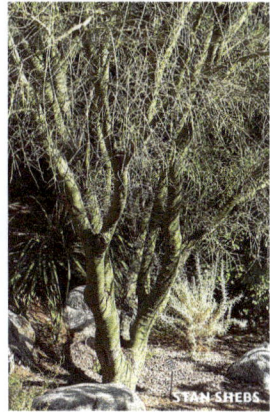

STAN SHEBS

PALOVERDE

Parkinsonia microphylla   **PEA FAMILY**

SYNONYM *Cercidium microphyllum*

Paloverde is Spanish for "green stick or tree," an excellent description for this low desert species. The tree stands 4.5–6 m (15–20 feet) high, and has smooth blue-green bark. The compound leaves are present only during spring, soon falling away with desert heat. The greenish bark then takes over the task formerly carried on by leaves. Flowers are typical of the pea family; seed pods are long and narrow with a constriction between each two seeds. Seeds have an especially hard coating that must be cracked or abraded before they will germinate. Native Americans produced a meal from the seeds by grinding them and mixing with water. Several kinds of desert wildlife use the seeds as food.

WHERE FOUND Primarily along the eastern shore of Lake Mohave, opposite Cottonwood Cove.

KRZYSZTOF ZIARNEK

LIDINE MIA

DON CARLSON

MESQUITE

Prosopis glandulosa   PEA FAMILY

ALSO CALLED Honey Mesquite

This many-branched shrub or small tree is one of the desert's valuable water indicating plants. Look for this handsome, thorny tree where springs occur, or water is near the surface of the ground. Its roots may penetrate 15–25 m (50–80 feet) in search of water. Flowers bloom in late April and May and attract many insects, especially honey bees. The fruits, resembling string beans, ripen in autumn, and are eaten by several kinds of animals. Native Americans ground the long, sweet pods into a meal. During pioneer days, the wood was used as fuel, fence posts, and in making furniture. Mesquite frequently acts as a windbreak, and wind-blown sand may pile up around a plant, almost burying it. It responds by sending out many more branches, while the trunk continues to grow underground. A thicket of mesquite is important habitat for small burrowing mammals as well as for birds and small reptiles.

STAN SHEBS

STAN SHEBS

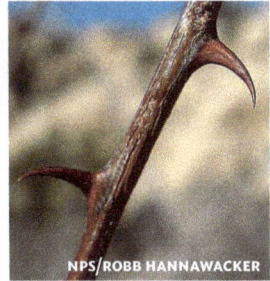

NPS/ROBB HANNAWACKER

CATCLAW

Senegalia greggii   **PEA FAMILY**

SYNONYM *Acacia greggii*

Numerous thorns, short and curved like a cat's claw, readily identify this deciduous shrub or small tree. The plant is heartily cursed by rider and hiker alike because of its ability to catch and tear clothes and flesh; for these reasons it is sometimes called "wait-a-minute" bush and "devil's-claw." The flower is very fragrant and an important source of honey. Seeds are in pods and were formerly used by Native Americans as food. Occasionally mistletoe causes large spindle-shaped swellings on this plant but is more common on mesquite. The thorny brunches offer an excellent nesting site for the verdin, while lower branches afford fine protection for small mammals.

WHERE FOUND Often forming thickets in the bottoms of canyons and washes.

JIM MOREFIELD

JOE DECRUYENAERE

HAIRY LOTUS

Acmispon strigosus   **PEA FAMILY**

ALSO CALLED Deer-Vetch

SYNONYM *Lotus tomentellus*

The main characteristic of this small prostrate annual is the compound leaves. The leaflets usually number from 4–8 and are covered with fine woolly hairs. Flowers are tiny, less than 8 mm (1/3 inch) long. The petals are often reddish on the back with age. Several seeds are produced and are gathered and stored by small rodents. The plant continues to grow as long as the soil remains moist.

WHERE FOUND In roadside gravels and along desert washes at all elevations of the southern portion of the region.

NPS/ ALESSANDRA PUIG-SANTANA

NPS/ ALESSANDRA PUIG-SANTANA

DESERT SENNA

Senna armata   PEA FAMILY

ALSO CALLED Desert Cassia

SYNONYM *Cassia armata*

This low, rounded, many-branched shrub adds little color of any kind to the desert during most of the year, which can appear dead and leafless much of the time. If winter rains furnish sufficient moisture, its smooth, pointed stems become greenish, a few small leaflets appear and the shrub bursts into a vivid mass of flowers in April and May. With the disappearance of the flowers, the tiny leaflets are shed to conserve moisture during the hot summer. Occasionally it receives enough late-summer rains to bloom again in the fall. The seeds are utilized for food by small mammals. WHERE FOUND Most conspicuous in desert washes or open areas throughout middle elevations of the region.

NPS

PATRICK ALEXANDER

HADLEY GARLAND

CREOSOTE-BUSH

Larrea tridentata   CREOSOTE-BUSH FAMILY

This is one of the most abundant plants of the Southwest deserts. The leaves often glisten as if freshly wet because of a coating of oily resin. This oil adaptation protects the plant from too great a water loss during the long, dry periods between rains. In April and May the plants take on a yellowish cast from the numerous flowers. After a few weeks, the shrub is covered with fluffy, white, ball-like fruit, each containing five seeds. Lac, a resinous incrustation on the branches, was used by Native Americans to repair pottery and glue arrowheads to arrow shafts, After a rain its name becomes evident, as it gives off a resiny, sweet odor. It is often referred to in Mexico by the name of "hediondillo," translated "little bad smeller" or "little stinker."

STAN SHEBS.

NPS/ANDREW CATTOIR

J. KEHOE

GLOBE-MALLOW

Sphaeralcea ambigua   MALLOW FAMILY

ALSO CALLED Desert-Mallow, Desert-Hollyhock

 A herbaceous perennial with a woody base, it grows as tall as 1 m (3 feet). The rounded leaves usually appear shrunken and wrinkled. Some people find the silvery hairs of the plant irritating to the eyes; in some parts of the Southwest they are called "sore-eye poppies." When in full bloom, the plant takes on a fiery glow in the late-afternoon sun. Globe-Mallow flowers throughout the year when conditions are right, but is most often seen from March to June.

WHERE FOUND Globemallows flaunt their graceful, blossom-covered stems on either side of the road between the Lake Mead Visitor Center and Las Vegas, and throughout most of the region, especially on disturbed soils and washes.

STAN SHEBS

NPS/ROBB HANNAWACKER

STAN SHEBS

ROCK-NETTLE

Eucnide urens   STICK-LEAF FAMILY

ALSO CALLED Stingbush

This plant is well named, as contact with it causes considerable discomfort. The skin soon develops a red, stinging rash similar to that from the familiar Stinging Nettle. Thousands of small, sharp-pointed poisonous hairs, and shorter barbed ones, give a silvery sheen to Rock-Nettle's leaves and stems. The leaves cling to clothing and are difficult to remove. The plant grows from 30–60 cm (1–2 feet) high. It flowers, creamy or pale yellow, bloom from April to June. Except for various insects and small reptiles, most animal life seems to shun the plant.

WHERE FOUND Prefers the rocky canyons under shady overhanging ledges, and is frequent along the shores of Lake Mead.

NPS/ANDREW CATTOIR

BLAZING STAR

Mentzelia tricuspis   BLAZINGSTAR FAMILY

ALSO CALLED Desert Corsage, Spiny-haired Blazing Star

This is a coarse-stemmed, much branched, annual with rough-textured stems and leaves which causes them to cling to clothing. It grows to a height of about 30 cm (1 foot) when it has sufficient moisture. The outer stamens of the flower are dilated and have three lobes at the end. The cream-colored flowers are in bloom from March to May.

WHERE FOUND A sun-loving plant, it is commonly found on south-facing slopes and low ridges at lower elevations.

JIM MOREFIELD

NPS/HALLIE LARSEN

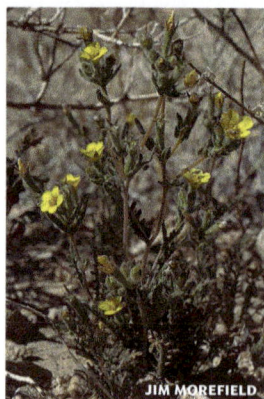

JIM MOREFIELD

SMALL-FLOWERED BLAZING STAR

Mentzelia albicaulis   BLAZINGSTAR FAMILY

This annual is branched at the base, with whitish, shining stems, and grows about 40 cm (15 inches) tall. Flowers are small and rather inconspicuous, measuring approximately 12 mm (1/2 inch) across, and appear from March to May.

WHERE FOUND It grows well in either the open or in the partial shade of shrubs, and is commonly found on sandy or gravelly soils of dry washes or in the open desert.

KRZYSZTOF ZIARNEK

WILLIAM HERRON

MOHAVE PRICKLY-PEAR

Opuntia phaeacantha   CACTUS FAMILY

The Mohave Prickly-Pear is primarily a spreading cactus with large, circular, flattened stems. Besides the noticeable curved spines, this plant also produces tiny hair-like spines called "glochids." Flower color, as in many cacti, is variable; in this species it ranges from bright yellow to peach. The fruits or "tunas" take on a purple color when mature and are commonly used in production of jellies, syrups, and wines. They also serve as food for many of the region's small desert mammals. Rabbits often feed on the petals during late evening hours.

WHERE FOUND This cactus has limited distribution in the Lake Mead area, visible only at higher elevations of the Newberry Mountains and Grand Wash Cliffs.

BUCKHORN CHOLLA

Cylindropuntia acanthocarpa   CACTUS FAMILY

ALSO CALLED Deerhorn Cholla, Cane Cholla

SYNONYM *Opuntia acanthocarpa*

This much branched, robust and long-stemmed cholla is common throughout the region. Typically, it achieves a height of 1.2 m (4 feet) or more. Flowers may be yellow, variegated, or tinged with red. Spines are 2.5 cm (1 inch) or more in length, stout, and quite numerous. Straight portions of the stems are often made into walking canes or other woodcrafts. Native Americans steam the buds and eat them in combination with pinole or salt-bush greens. The fruits are also eaten by rodents, such as the woodrat.

WHERE FOUND Open basins and valleys in the Creosote-Bush Community.

STAN SHEBS

STAN SHEBS

HOLY CROSS CHOLLA

Cylindropuntia ramosissima   **CACTUS FAMILY**

ALSO CALLED Darning-Needle Cactus, Diamond Cactus, Pencil Cholla
SYNONYM *Opuntia ramosissima*

This is an easy cactus to miss, as it often grows in the shadow of larger desert shrubs. It is a much-branched plant with very slender, woody stems. The spines are long and formidable in appearance. As with most chollas, the long central spines are encased in a sheath, like a sword in its scabbard. Flowers, when present, are small and inconspicuous, hence often overlooked. The plant is most often noticed in late-fall and winter because of the handsome 2.5 cm (1 inch) fruits. Old stems lose their spines and appear reddish in color in late summer.

WHERE FOUND Prefers desert flats and is common along the roads to Temple Bar and Davis Dam.

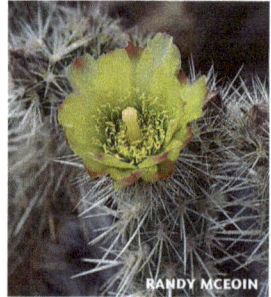

SILVER CHOLLA

Cylindropuntia echinocarpa   **CACTUS FAMILY**

ALSO CALLED Straw-Top Cactus

SYNONYM *Opuntia echinocarpa*

This cactus grows as a low shrub with a few ascending stems and many short, spreading branches The numerous straw-colored spines, borne on short tubercles, are covered with thin, papery sheaths. When these are moistened they taste like witch-hazel. It blooms in April and May. The flower petals are sometimes tinged with red on the outer petals. Flowers are followed by dry, spiny fruits that often do not mature. The Silver Cholla seldom forms thickets. After it has dried, the woody tubular stem skeletons, appropriately called "ventilated wood" by desert campers, make excellent fuel.

WHERE FOUND Low elevations, preferring wide gravelly washes and open slopes.

STAN SHEBS

STAN SHEBS

OLD MAN PRICKLY-PEAR

Opuntia polyacantha   **CACTUS FAMILY**

ALSO CALLED Grizzly-Bear Cactus

SYNONYM *Opuntia erinacea*

The "old man" cactus received its name from the dense white or pale gray beard-like covering of spines. Flowers are bright yellow, often tinged with pink or red, or sometimes may be a rich reddish-purple. The pad-like joints form mats that may rise 60 cm (2 feet) or more in height.

WHERE FOUND Prefers gravelly slopes in higher elevations and may be found in the Newberry Mountains. A reddish-flowered variety grows in the Pearce Ferry area associated with the Joshua-Tree forest. Both varieties bloom from late April to June.

STAN SHEBS

JOE DECRUYENAERE

DORNENWOLF

COTTON-TOP CACTUS

Echinocactus polycephalus   CACTUS FAMILY

The name "cotton-top" refers to the generous tufts of white cottony hairs enveloping the flower base and fruit. Because these "balls of cotton" are present throughout most of the year, species recognition is easily accomplished. The stems are usually clumped, often in groups of more than a dozen, and grow only 30 cm (1 foot) or so in height. It prefers the gravelly slopes and is seldom found elsewhere. Spines and flowers resemble the Solitary Barrel Cactus, but their distribution differs.

WHERE FOUND Cotton-Top Cactus ranges from the Creosote-Bush Community to the Pinyon and Juniper Community. They may be seen in several places in the recreation area, but especially as one descends from Searchlight to Cottonwood Cove and around Pearce Ferry.

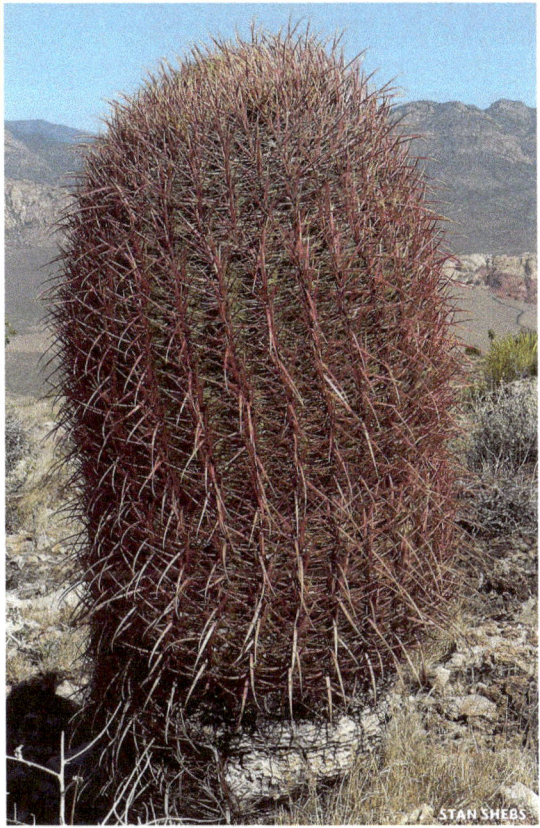

SOLITARY BARREL CACTUS

Ferocactus cylindraceus   **CACTUS FAMILY**

ALSO CALLED Bisnaga

SYNONYM *Ferocactus acanthodes*

Together with other species, this cactus is also called "bisnaga." The Solitary Barrel Cactus, at first globular in form, eventually becomes cylindrical and may grow to a height of 1.5–1.8 m (5–6 feet). It tends to grow toward the sun, which often produces a leaning habit frequently seen in older plants. The flat reddish spines exhibit rings approximately 3 mm (1/8 inch) apart. Flowers are yellow and arranged in a circular pattern at the apex of the stem. While it has been called a useful thirst quencher, the slimy alkaline juice obtained by mashing the pulpy flesh might in an emergency save a life, but is not suitable to drink. Woodrats have been known to eat into the plant at the base, living off of the soft tissues and water while being well protected from enemies.

NPS/HANNAH SCHWALBE

STAN SHEBS

SUNDROP

Chylismia brevipes EVENING-PRIMROSE FAMILY

ALSO CALLED Yellow Cups, Golden Evening-Primrose

SYNONYM *Camissonia brevipes*

This is one of the most noticeable of several different kinds of primroses in the region. It may grow to a height of 60 cm (2 feet) and produce from 1–6 vertical stems. Each stem may have several flowers, with each flower about 2.5 cm (1 inch) across. The toothed leaves are mostly confined to a basal rosette, usually emerging in early spring or late winter months. The leaves are often red-veined beneath.

WHERE FOUND Sundrop is an abundant plant of broad gravel washes and rolling hills, blooming from March to May, and is an important food for wildlife, especially desert bighorn sheep.

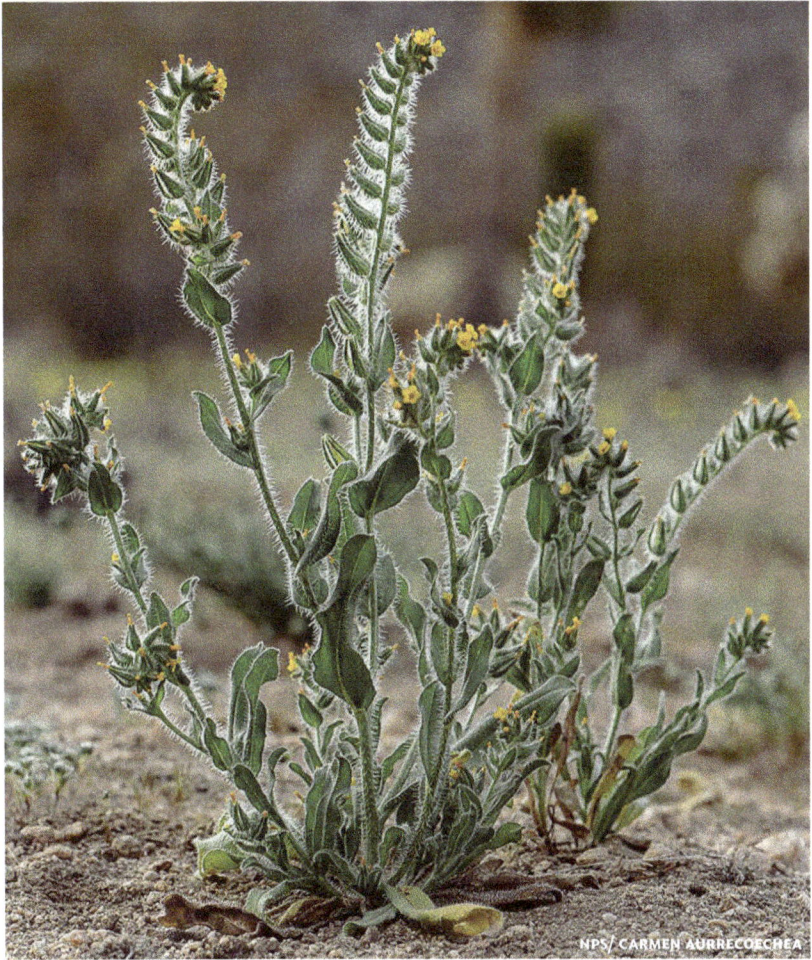

NPS/ CARMEN AURRECOECHEA

FIDDLENECK

Amsinckia tessellata   **BORAGE FAMILY**

It is thought by some that the forget-me-nots (including this species) received their name from the dense covering of clear, stiff, prickly hairs. Once
you've become acquainted with these plants you, too, will not forget them.
Fiddleneck possesses small flowers less than 1.2 cm (1/2 inch) in length. The
arrangement of these flowers in one-sided coils resembles the structure of
the neck of a violin, hence the common name. Each flower produces a nutlet
fruit composed of four parts. Seeds are believed to cause digestive problems
in grazing cattle and sheep.

WHERE FOUND Fiddlenecks are abundant in the open Creosote-Bush Community, especially in southernmost portions of the Lake Mead recreation area.

STAN SHEBS

DICK CULBERT

GROUND CHERRY

Physalis crassifolia   NIGHTSHADE FAMILY

This is a rather compact plant and a common desert perennial. It is usually low-growing, but under favorable moisture conditions may grow to a height of over 30 cm (1 foot). The flower is pendant and saucer-shaped when open. After blooming (from April to July), papery seed pods are formed, with a many-seeded berry inside. It is related to the tomato, and the berries are edible and may be eaten raw or cooked. Some people have used them in making preserves. Native Americans valued the berries for food. Several small animals find them an attractive food.

WHERE FOUND In desert washes and around protective boulders.

STAN SHEBS

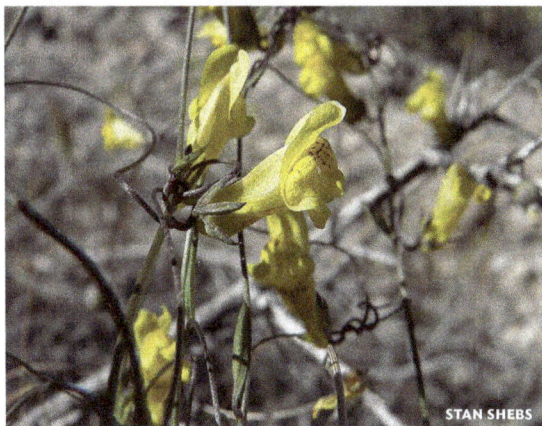

STAN SHEBS

TWINING SNAPDRAGON

Neogaerrhinum filipes   **PLANTAIN FAMILY**

SYNONYM *Antirrhinum filipes*

This flower resembles the garden snapdragon, but is smaller. It grows on a bright green, thread-like stem which climbs by twisting around the low branches of desert shrubs, thus receiving only limited exposure to sunlight. It blooms in April and May.

WHERE FOUND Never found in any great abundance, Twining Snapdragon is confined to the Creosote-Bush Community where it is often associated with Bur-Sage growing in the washes.

STAN SHEBS

LESSER MOHAVEA

Mohavea breviflora   **PLANTAIN FAMILY**

The flowers of Lesser Mohavea resemble the familiar garden snapdragon, and are partially hidden by the leafy stem. The plant grows only about 15–20 cm (6–8 inches) in height, and its stem is branched from the base. It is covered with many glandular hairs, resulting in a very sticky texture. Because of this, sand particles are usually found adhering to the stem and leaves. A short-lived annual, it germinates, flowers, and produces seed in a three-week period.

WHERE FOUND Lesser Mohavea is a plant of lower elevations, and prefers the steep talus slopes of lava hillsides, but also occurs on gravelly banks of washes.

COYOTE-MELON

Cucurbita palmata   GOURD FAMILY

Like other gourds, this plant spreads out for several feet, with stems being reaching as long as 3 m (10 feet). Leaves are large, palmately lobed, and dark green. The plant usually has both male and female flowers. The female flower is large, with a 5-lobed pistil; the male flower is smaller, the anthers formed into a single central knob. The mature gourd is yellow and up to 10 cm (4 inches) or more in diameter. The gourd is a favorite food of the coyote, hence the common name.

WHERE FOUND Prefers gravelly banks of washes and is frequently seen growing in disturbed soils along roadsides.

MATT LAVIN

MATT LAVIN

BROOM SNAKEWEED

Gutierrezia sarothrae   **SUNFLOWER FAMILY**

ALSO CALLED Matchweed, Snakeweed

These small, rounded shrubs bloom during late summer and fall, and often occur in almost pure stands. The flowering heads are small but quite numerous, sometimes numbering into the hundreds on a single shrub. They are clustered in bunches at the ends of the many branches. The leaves are resinous and burn readily when dry. The flowers are a favorite with bees.

WHERE FOUND Matchweed grows best on open slopes and is especially abundant around 910 m (3,000 feet) or more in the Joshua Forest Community.

STAN SHEBS

PATRICK ALEXANDER

PATRICK ALEXANDER

SPINY GOLDENBUSH

Xanthisma spinulosum   SUNFLOWER FAMILY

SYNONYM *Haplopappus gooddingii*

It is a perennial, with several leafy stems arising from a woody base, It grows 15–40 cm (6–16 inches) tall and is bright green. The leaves are short and slender with many bristle tipped lobes. Its green foliage and seeds furnish food for small rodents.

WHERE FOUND This plant seems to grow best in gravelly areas and on rocky canyon walls. Look for it in late summer and fall.

GOLDEN-EYE

Viguiera parishii   **SUNFLOWER FAMILY**

ALSO CALLED Nevada Viguiera, Resin Weed

SYNONYM *Bahiopsis parishii, Viguiera deltoidea*

This low shrub looks very similar to the Rayless Encelia, and the two are sometimes found growing in the some area. It grows to a height of 30–60 cm (1–2 feet). The lower stem leaves are opposite, strongly toothed, heavily veined, with edges rolled under. The medium-sized flower heads are numerous, growing on the tips of long, slender stems, either solitary or in small flat-topped clusters. Blooms in late summer and early fall.

WHERE FOUND Prefers open gravelly slopes; common at higher elevations, and especially abundant near Christmas Tree Pass in the Newberry Mountains.

NPS/ANDREW CATTOIR

STAN SHEBS

NPS/ANDREW CATTOIR

SUNRAY

Enceliopsis argophylla   SUNFLOWER FAMILY

Sunray is one of the most impressive members of the Sunflower family to be found at Lake Mead. The large flowers, as much as 15 cm (6 inches) across and on leafless stems, rise a foot or more above basal tufts of large, silvery-gray leaves. The leaves are rather thick, smooth, covered with a fine down, and feel similar to felt. A closely related variety, almost identical in appearance, is the Panamint Daisy (*Enceliopsis covillei*) found only in Death Valley National Park.

WHERE FOUND Partial to eroded soils containing gypsum, Sunray is especially noticeable along the North Shore Road from Las Vegas Wash to Overton, and in the Kingman Wash and Bonelli Landing areas.

CURTIS CLARK

CURTIS CLARK

FOREST AND KIM STARR

BRITTLE-BUSH

Encelia farinosa   **SUNFLOWER FAMILY**

ALSO CALLED Incienso, Incense Bush

Flowering seven months of the year, Brittle-Bush exhibits all the characteristics of a well-adapted desert shrub. The plant stands as much as 1 m (3 feet) tall and is covered with numerous large, silvery leaves. During periods of high temperatures the leaves may dry and fall away, leaving only naked stems. These stems were dried and burned as incense in the missions by early Spanish padres.

WHERE FOUND Able to survive in the most extreme localities, it is most common on barren, south-facing slopes, and in rocky areas.

CURTIS CLARK

STAN SHEBS

RAYLESS ENCELIA

Encelia frutescens   SUNFLOWER FAMILY

This rounded, much branched shrub is similar in appearance to its relatives, the Encelia and Brittle-Bush. Its leaves are green, without any coating, and are scattered along the whitish stems. The plant has only disk flowers.

WHERE FOUND Prefers dry slopes and desert flats in the Creosote-Bush Community. Commonly found at elevations up to 1,220 m (4,000 feet), it is especially noticeable in the Newberry Mountains near Davis Dam.

STAN SHEBS

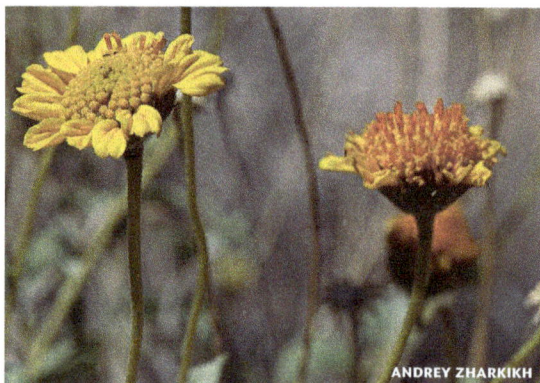

ANDREY ZHARKIKH

ENCELIA

Encelia virginensis   **SUNFLOWER FAMILY**

This shrubby plant closely resembles the Rayless Encelia, but differs in that a coating of fine, grayish-white hairs cover its green leaves, a feature lacking in Rayless Encelia. Each flower stalk bears a single head, having both ray and disk flowers. It may bloom during both spring and fall months. It produces quantities of seeds which are stored and eaten by small rodents. Some Indian tribes boiled the leaves and flowers, and used the reulting brew as a wash to relieve rheumatism.

WHERE FOUND On the banks of rocky washes and ridges at middle and higher elevations of the desert.

MATT LAVIN

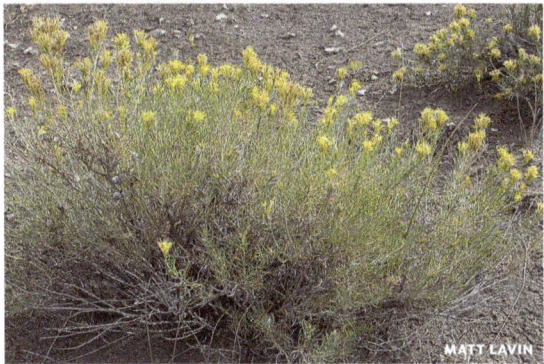

MATT LAVIN

RABBITBRUSH

Ericameria nauseosa   SUNFLOWER FAMILY

SYNONYM *Chrysothamnus nauseosus*

This shrub is 1 m (three feet) or more in height, and composed of a large number of stems; the stems are clothed with a felt-like covering. Leaves are gray-green and very bitter to the taste. The plant may become leafless during high summer temperatures to reduce moisture loss from transpiration. Flowers are small but numerous, giving ends of the branches a rich yellow cast. The rayless flowers are borne in clusters. In spite of its name of "nauseosus," the plant has a pleasant odor. The blooming of Rabbitbrush heralds the approach of fall. Several other species of Rabbitbrush, somewhat similar in appearance to this one, are found in the region

WHERE FOUND In washes and around slopes.

STAN SHEBS

ALAN VERNON

STAN SHEBS

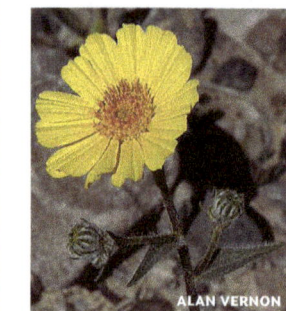

DESERT-SUNFLOWER

Geraea canescens   **SUNFLOWER FAMILY**

ALSO CALLED Desert Gold

These plants often form vast gardens of color on sandy desert basins and along roadsides. Even when there isn't much rainfall in the cooler months to furnish needed moisture, these annuals seem to bloom anyway. Flowers begin to show as early as February at lower elevations and continue through May. Occasionally conditions are right for another flower season in early fall. Plants are large and showy, about 60 cm (2 feet) high, with many short white hairs on leaves and stems, and is common throughout the area. Many small rodents use the seeds as part of their diet.

NPS/ANDREW CATTOIR

NPS/ANDREW CATTOIR

SWEETBUSH

Bebbia juncea   SUNFLOWER FAMILY

ALSO CALLED Chuckwalla's Delight

Leafless most of the year, this shrub is well-adapted to the desert. The round-ed plant often reaches a diameter of 1.2–1.5 m (4–5 feet). Stems are numer-ous and rough to the touch. The orange flowers are small, but numerous, in the late spring. When conditions are right, it may bloom again in the fall. As the common name indicates, it is a favorite food for the chuckwalla lizard. It is also eaten by desert bighorn and visited extensively by bees.

WHERE FOUND Common in broad sand and gravel washes.

STAN SHEBS

CHRIS ENGLISH

DESERT-MARIGOLD

Baileya multiradiata   SUNFLOWER FAMILY

ALSO CALLED Wild Marigold

Often producing the "yellow lining" along the roadsides, Desert-Marigold blooms throughout much of the year. The flowers, one to each stem, are 2.5 cm (1 inch) in diameter and positioned at the ends of leafless stems. The stem bases are somewhat woody; the leaves are mostly basal and downy. It produces large numbers of seeds, an important food for small desert rodents. WHERE FOUND Prefers well-drained, rocky or gravelly slopes, and is abundant in disturbed soil along road shoulders throughout the region.

FOREST AND KIM STARR

FOREST AND KIM STARR

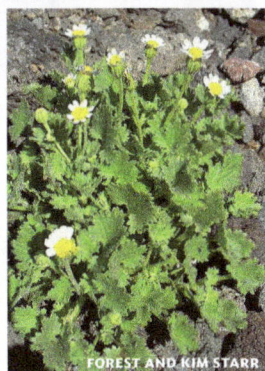
FOREST AND KIM STARR

ROCK DAISY

Perityle emoryi   SUNFLOWER FAMILY

This is a fragile, gland-covered annual with a strong scent when crushed. The leaves are large and showy, with a cool-clammy texture. Flowers are small and not especially noticeable. The roots are typically found under large boulders, tapping the moisture captured there.

WHERE FOUND Because of its preferred habitat, it is properly named—it is most common among boulders of steep talus slopes and around the base of cliffs during the period from April to June (although it may also become abundant in early fall if sufficient moisture is available).

STAN SHEBS

PATRICK ALEXANDER

ANDREY ZHARKIKH

PAPER-FLOWER

Psilostrophe cooperi **SUNFLOWER FAMILY**

This short, rounded shrub stands 30–45 cm (12–18 inches) high and is woody only at its base. It has many stems covered with a white woolly down and simple undivided leaves 2.5–7.5 cm (1–3 inches) in length. In a strong breeze, the rustling of dried flowers resembles the warning sound of a rattlesnake, somewhat disconcerting to the hiker! While it blooms in the spring months, it may have a second flowering in early fall if there have been sufficient late-summer rains.

WHERE FOUND Prefers sandy and gravelly areas at low to middle elevations.

STAN SHEBS

NPS/ANDREW CATTOIR
WOOLLY ERIOPHYLLUM

STAN SHEBS
WOOLLY-DAISY

WOOLLY-DAISY

Eriophyllum wallacei   **SUNFLOWER FAMILY**

ALSO CALLED Wallace Eriophyllum

Another small annual which can be classified as a "belly flower," Woolly-Daisy is less than 5 cm (two inches) tall and is covered with numerous soft hairs. A single plant may produce as many as a dozen or more flower heads. These plants may become abundant, producing a fluffy carpet following periods of above-average rainfall. Flowering season is April and May. Although small, the flowers produce quantities of seeds which are eaten by small rodents. A closely related species found in the same region, differing mainly in having white flowers, is the Woolly Eriophyllum (*Eriophyllum lanosum*).

WHERE FOUND In rocky soils and broad gravel and sandy washes.

STAN SHEBS

YELLOW-HEAD

Trichoptilium incisum   **SUNFLOWER FAMILY**

ALSO CALLED Yellow Pincushion

If sufficient moisture is available at the right times, this aromatic annual may bloom between March and May and again late in September and October. The round, ball-like flower head is made up of numerous small, tube-shaped flowers. These heads are borne on slender, reddish stems 7.5–10 cm (3–4 inches) in length. It is a low-growing plant with the stem and toothed leaves covered by hairlike wool.

WHERE FOUND Yellow-head grows best in open areas of desert pavement in the Creosote-Bush Community.

STAN SHEBS

JIM MOREFIELD

CHINCH-WEED

Pectis papposa   **SUNFLOWER FAMILY**

This fall-flowering annual is a conspicuous feature of the desert's autumn color. It possesses a strong, but pleasant, odor associated with the small, dark-covered glands at the base of the flower and on some of the leaves. Following adequate summer rains, it may provide a yellow hue to blend in with the desert browns and greens.

WHERE FOUND Look for this plant in September and October throughout low to middle elevations of the region.

STAN SHEBS

TYRRHIUM

JIM MOREFIELD

DESERT-FIR

Peucephyllum schottii   **SUNFLOWER FAMILY**

ALSO CALLED Pygmy-Cedar, Sprucebush

This shrub received its several names because its vivid green, needle-like leaves and general shape give a superficial resemblance to a cone-bearing tree. In reality, its flowers reveal membership in the Sunflower family. It may attain a height of 1.5 m (5 feet), but is usually smaller. The resinous leaves emit an aromatic fragrance when crushed. Desert-Fir blooms from early spring to May.

WHERE FOUND Along steep rocky slopes and in many dry washes. It is especially common in the Black Canyon area.

STAN SHEBS

DAWN ENDICO

EMILIE GRENIER

TURTLEBACK

Psathyrotes ramosissima   SUNFLOWER FAMILY

ALSO CALLED Desert-Rosette

The growth form resembles the back of a turtle, hence the common name. It is a very compact plant, with thick, gray-green, wrinkled leaves that are covered with numerous hairs. Crushed leaves have a strong turpentine odor. Stems are very brittle. The flowers are small and rather inconspicuous.

WHERE FOUND Prefers gravelly areas, especially where the soil has been disturbed along roadsides. It is common in the Las Vegas Wash area along the North Shore Road.

PATRICK ALEXANDER

STAN SHEBS

STAN SHEBS

GROUNDSEL

Senecio flaccidus   **SUNFLOWER FAMILY**

ALSO CALLED Sand-Wash Groundsel

SYNONYM *Senecio douglasii*

This plant may grow as much as 1 m (3 feet) high. The stems are somewhat woody at the base, erect, and bear thread-like leaves and large showy flowers. The flowers frequently have an unusually long blooming season. While the season is usually from March to May, a second period may occur in the fall.

WHERE FOUND Groundsel is most often found in desert washes in Joshua forests and low mountain ranges.

ANDREY ZHARKIKH

YELLOW SAUCERS

Malacothrix sonchoides   SUNFLOWER FAMILY

This colorful annual grows up to 30 cm (1 foot) in height. The large, deeply lobed leaves are mostly basal, with a few smaller leaves along the stems. The flower heads are fragrant, and several may be found on one plant. It is closely related to the Desert-Dandelion. Although the two plants are very similar, they are usually not found growing together. It is in sandy areas that Yellow Saucers have become adapted, and are quite common in some years. An abundance of seeds is produced, and these are much relished by small rodents and seed-eating birds.

STAN SHEBS

JAREK TUSZYNSKI

ROCKPOCKET

DESERT-DANDELION

Malacothrix glabrata   SUNFLOWER FAMILY

The flower heads of this plant look very much like the familiar dandelion so common in lawns, with one difference—the center of the desert variety has a small reddish "button." Depending on rainfall, it may produce numerous blossoms or just a few. In good flower years, the desert is carpeted in many areas with these plants. Its green leaves furnish food for several kinds of wildlife, and the flower heads are eagerly eaten by the desert tortoise.

WHERE FOUND In open areas and on rolling hills.

NPS/ANDREW CATTOIR

MATT BERGER

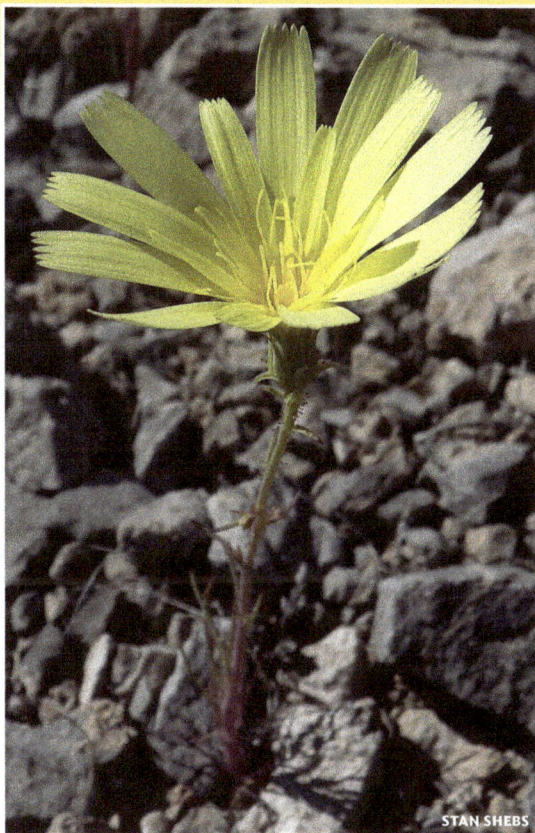
STAN SHEBS

PARRY TACKSTEM

Calycoseris parryi   SUNFLOWER FAMILY

ALSO CALLED Yellow Tackstem

Two closely related tackstems occur in the Lake Mead region (see **White Tackstem**). The only noticeable difference is in flower color. The white variety is the more common of the two. Like its white cousin, the yellow receives its name from the small but numerous dark, tack-shaped glands which cover upper portions of the stem.

WHERE FOUND On open gravelly areas and rolling hills.

ANDREY ZHARKIKH

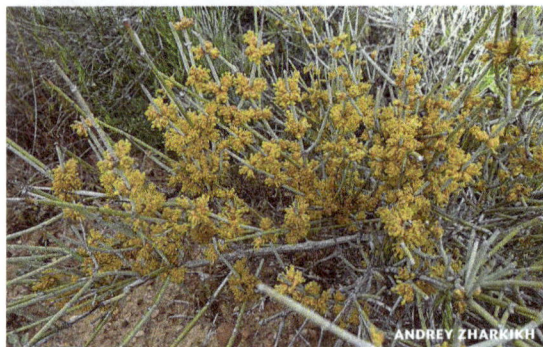

ANDREY ZHARKIKH

MORMON-TEA

Ephedra nevadensis   MORMON-TEA FAMILY

ALSO CALLED Desert-Tea, Mexican-Tea, Joint-fir

Sometimes called "joint-fir" because of its jointed stems, this plant is closely related to the pines, cycads and ginkos. The stems bear scale-like leaves, and develop small yellowish cones at joints on the stem. It was commonly used by Native Americans as a brew and a medicine, and later by Mormon pioneers as a refreshing tea. The brew is somewhat laxative. The plant contains a high concentration of tannin, which helps explain its early use as a medicinal tonic and poultice. Its small black seeds were roasted and eaten. Birds and small mammals, especially antelope ground squirrels, harvest them in great quantity.

WHERE FOUND Mormom-Tea grows best in washes and on higher desert slopes.

STAN SHEBS

NPS/ROBB HANNAWACKER

DESERT MISTLETOE

Phoradendron californicum   SANDALWOOD FAMILY

The mistletoe is a parasite; that is, it survives at the expense of its host. Large clumps are commonly found growing on Honey Mesquite and Catclaw, but may sometimes be found on Creosote-Bush, Paloverde, and Saltcedar. Branches of these plants may suffer from heavy infestations of this parasite. The flowers are fragrant and obscure. Red berries appear during summer and remain into the fall. Birds, such as the phainopepla, bluebird, and Gambel's quail are particularly attracted to the berries as a food source and, consequently, assist in seed dispersal. The juicy pulp of the berries helps supply water as well as food for wildlife during dry summer months.

NPS/ANDREW CATTOIR

MANFRED WERNER

STAN SHEBS

DESERT-HOLLY

Atriplex hymenelytra   AMARANTH FAMILY

This silvery shrub is another desert plant found on gypsum-rich soil. Leaves are thick, somewhat leathery to the touch, and covered with a velvety down. Later in summer the leaves take on a pinkish tinge and many tend to drop off, especially if rainfall is unusually sparse. Flowers are inconspicuous and wind-pollinated, thus easy to miss. The seeds are flat, papery and rounded, about the size of oatmeal flakes. In some parts of the desert this plant has been collected until it is quite rare. The silvery leaves make it a desirable Christmas decoration, and in the past (before being legally protected), quantities have been sold for that purpose.

WHERE FOUND Common along the North Shore Road.

JIM MOREFIELD

JASON HOLLINGER

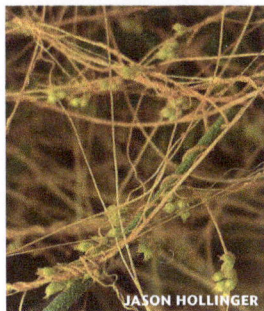

JASON HOLLINGER

DODDER

Cuscuta californica   **MORNING-GLORY FAMILY**

What appears to be a network of orange thread entangled within the branches of desert shrubs is actually a twining parasitic plant. The seeds germinate in the soil, but quickly send up frogile stems, attaching to the woody tissue of the shrub. Runners from the stem penetrate the host and result in dodder losing all connection with the soil, thus becoming a true parasite. Petals are absent.

WHERE FOUND In large desert washes, with Hemenway Wash a preferred locality.

JIM MOREFIELD

STAN SHEBS

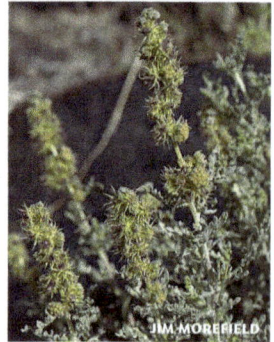

JIM MOREFIELD

BUR-SAGE

Ambrosia dumosa   SUNFLOWER FAMILY

ALSO CALLED Burro-Bush

Bur-Sage is one of the most common and widespread desert shrubs in the lands surrounding Lake Mead. The flowers are small, inconspicuous, without petals, and are wind pollinated. Only during the spring are its leaves green. As summer heat arrives, the leaves lose their color, become ashy-white, and the entire plant appears to be dead or dying. When crushed, the leaves give off a pleasant sage-like odor. In spite of the plant's bitter taste, it is one of the important foods of the desert bighorn, wild burro, and some of the smaller animals.

WHERE FOUND Most often associated with Creosote-Bush. It prefers open desert flats and slopes of dry washes.

JIM MOREFIELD

ANDREY ZHARKIKH

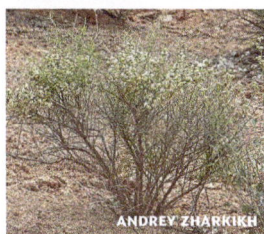
ANDREY ZHARKIKH

CHEESE-BUSH

Ambrosia salsola   SUNFLOWER FAMILY

ALSO CALLED Bush-Hops, White Burro-bush

SYNONYM *Hymenoclea salsola*

Called "cheese-bush" because the foliage, if crushed, has a cheese-like odor. Its leaves are thread-like, 2.5–5 cm (1–2 inches) in length, bright green, and resinous. From a distance, the plant may be mistaken for the weedy Russian Thistle (*Kali tragus*). The flowers are inconspicuous, but the fruits are interesting and can be showy when in great number. Look for the white or pink-tinged scales that surround the young fruit like a small rosebud. During hot summer months the plant appears to be made up of dead stems.

WHERE FOUND Cheese-Bush is one of the characteristic plants of the Desert Wash Community and is abundant in Hemenway Wash.

SELECTED REFERENCES

Baldwin Bruce G., et al. 2002. *The Jepson Desert Manual: Vascular Plants of Southeastern California.* University of California Press.

Chadde, Steve. 2019 *Southwestern Trees: A Guide to the Trees of Arizona, New Mexico, and the Southwestern United States.* Orchard Innovations.

Chadde, Steve and Roxana Ferris. 2020. *Death Valley Wildflowers: A Visitor's Guide to the Wildflowers, Shrubs and Trees of Death Valley National Park.* Orchard Innovations.

Jaeger, Edmund C. 1940. *Desert Wild Flowers.* Stanford University Press.

Mackay, Pam. 2013. *Mojave Desert Wildflowers: A Field Guide To Wildflowers, Trees, and Shrubs of thhe Mojave Desert, including thhe Mojave National Preserve, Death Valley National Park, and Joshua Tree National Park.* Falcon Guides.

Munz, Philip A. 2004. *Introduction to California Desert Wildflowers.* California Natural History Guides, Volume 74. University of California Press.

ACKNOWLEDGMENTS

The author extends his sincere gratitude to a number of photographers whose work was used extensively in this book (under the appropriate Creative Commons license): Stan Shebs, Matt Lavin, Patrick Alexander, Andrey Zharkikh, Jim Morefield, Forest and Kim Starr, and several National Park Service photographers.

INDEX

NOTE: Synonyms are listed in *italics*.